Heredity and Humanity: Race, Eugenics and Modern Science

Heredity and Humanity:

Race, Eugenics and Modern Science

Roger Pearson

Scott-Townsend Publishers
P.O. Box 34143, Washington D.C. 20043

ISBN 1-878365-15-5

Author Roger Pearson was born in England in 1927 and received his Master's degree in Economics and Sociology and his Ph.D. in Anthropology from the University of London. Leaving England he served as an officer in the British Indian Army and spent sixteen years in Asia and Africa before coming to the United States in 1965, where he was department head of anthropology in two U.S. universities and dean of academic affairs and director of research at a third. Amongst other works he is the author of *Eastern Interlude* (Thacker Spink, Calcutta) and several textbooks, including *Introduction to Anthropology* (Holt, Rinehart and Winston) and *Anthropological Glossary* (Krieger Publishing Company), as well as *Race, Intelligence and Bias in Academe* (Introduction by Arthur R. Jensen) and *Shockley on Eugenics and Race* (Introduction by Hans J. Eysenck), both published by Scott-Townsend. The last three books are still in print. He is currently working on *A Modern Guide to Anthropology*.

Scott-Townsend Publishers
P.O. Box 34143, Washington D.C. 20043
Tel: (202) 371-2700; Fax: (202) 371-1523

To the memory of all those young men whose death without heirs in World Wars I and II lowered the quality of the gene pool of the West

Contents

I

THE CONCEPT OF HEREDITY IN THE ANCIENT WORLD

THE GENERAL reading public, including a high percentage of those who have been exposed to contemporary university courses in the humanities, fail to appreciate the significant role played by the concepts of race and heredity in the history of Western thought. A brief outline of this history, showing the commitment of the Western world to a recognition of the efficacy of heredity – until academic and media attitudes were affected in the first half of the present century by changes in the social, political and demographic climate – is therefore of value. This chapter illustrates the deep belief in the importance of heredity and race which prevailed from the earliest times until roughly the end of the nineteenth century. Subsequent chapters document the rise of politically-motivated egalitarian ideology which, heavily supported by the media, eventually succeeded in making the idea of biological *inequality* taboo. Despite the fact that there is today a rapidly developing body of scientific research which validates the age-old comprehension of the role of heredity in shaping human abilities, too many people are unaware of the mechanics behind the swing toward the notion of the *biological equality* of mankind. It is to be hoped that the following observations will be enlightening.

Western tradition has long recognized that heredity plays a significant role in determining not merely the characteristics of plants and animals but also the mental and physical qualities of human beings. Some elementary recognition of the role of genetics as a causal force may have originated as early as the Neolithic revolution, when cultivators learned how to improve upon the various species of wild grasses and to breed domesticated milk- and meat-providing animals which were more useful to mankind than those they found in the wild. By the time of the great classical civilizations of ancient Greece and Rome it had become commonplace knowledge – based on observing and remembering the generations from the same family – that heredity also played an important role in determining the character and abilities of men and women.

In most early European societies, as in virtually all early societies that achieved an advanced culture, the social group was seen by its members as an intergenerational affair, with the family and the ancestors playing an important role in the self-concept of the individual. In early times the term "race," or its equivalent in other European languages, was loosely used to refer to any lineage or group of people descended from common ancestors – a useful work-term far more practicable than the narrow definitions attempted in recent times. Life does not begin, nor does it end, with the individual. As Denis Fustel de Coulanges pointed out in 1864, in his classic study of ancient Greek and Roman culture entitled *The Ancient City* (1955), it was the idea of common descent from the same ancestral forebears – the idea of belonging to a specific community of families, and of sharing the same, hopefully eternal, thread of life – that held the freemen of the Greek city-state together. As long as the lineage survived, the ancestors lived on in the minds and bodies of their des-cendants; death was only final when the entire nation was eliminated. The biological reality was interpreted into religious terms. The individual was seen as the product of the forces of biological causality, a living link in the chain that was the lineage, just as the lineage comprised a vital component of the nation-state, and the nation-state was a distinctly biological unit, with its own distinctive gene pool:

> Reproduction in the ancient community was a religious duty... The religious society was the family, the genos.

> Paternal dignity and sacerdotal dignity were fused: the eldest son, upon the death of the father, becomes the head and priest of the family. The deceased father is honored by his children as a kind of divinity. He himself is honored by his children as a kind of divinity. He himself rendered the same worship to his ancestors: thus the greatest misfortune that his piety had to fear, is that the line shall be stopped. For then his religion would disappear from the earth, his hearth would become extinct, the whole series of his departed ones would fall into oblivion ...

The qualities that characterized individuals were acquired, it was believed, from their ancestors. Thus we find a speaker in the *Odyssey* (*IV*, 60) observing that "the blood of your parents was not lost in you, but ye are of the line of men that are sceptered kings, the fosterlings of Zeus, for no churl could beget sons like you." Similarly there are references to the disguised Athena as being "delicate of countenance such as are the sons of kings" (*XIII*, 216), whereas in the *Iliad* Thersites is described as ill-formed with a warped head. It was recognized that even well-born individuals had to be schooled and trained to develop their inborn qualities to the maximum, but basic potential was inborn. In Homeric Greece, even truthfulness – a revered value – was deemed to be an inherited virtue, and to call a *eupatrid,* or "person of good ancestry," a liar was tantamount to calling him a bastard, a man of impure, inferior descent. Indeed, we find personalities in the *Iliad* making statements such as: "Therefore ye could not say that I am weak and a coward by lineage, and so dishonor my spoken word" (Il. *XIV*, 126). Even as late as the age of Classical Athens, Aristotle described the physical and moral characteristics that were deemed to constitute nobility as being "an inherited virtue" (Pol. *IV*. 8).

According to L. R. Palmer, the authority on the Pylos tablets, Achaean kings held their office by virtue of the purity of their descent. Among the Achaeans, he wrote: "Where the 'luck' of the tribe is concerned, there is no substitute for blue blood" (1955, p. 9). Werner Jaeger went even further, describing the Hellenic ideal as an "aristocracy of race (1945, p. 205)."

Because of their respect for good breeding, the Greeks honored their women as the progenitors of the race, and it was said that men chose their wives as they chose their horses, by the length of their pedigrees. The desirability of breeding from proven stock had become a cultural requirement, and only children born of legitimate wives (i.e., of quality ancestry) could inherit the social status of the father. Indeed, in ancient Athens and other Greek city-states, the eupatrids were men descended from no less than nine generations of untainted noble stock on both sides of the family tree.

Plato's interest in eugenics is well known, and he praises the Spartans for their eugenic breeding policies (*Laws*, 630). Aristotle is equally impressed by the need to breed good stock. Theognis of Megara constantly praises the importance of heredity, lamenting that "[W]e seek well-bred rams and sheep and horses and one wishes to breed from these ... [but] men revere money, and the good marry the evil, and the evil the good. Wealth has confounded race." (*Theognis*, *V.* 183). Racial purity was linked to physical appearance, with Spartan women being renowned for their beauty; and character was seen as inherited along with personal features: "Thou art pleasing to look upon and thy character is like to thy form" (*Stobaeus*, lxxxviii, 71). In Greek literature the importance of heredity is repeated again and again: "Noble children are born from noble sires, the base are like in nature their father" (*Alcmeaon*, *Fr.* 7); "I bid all mortals beget well-born children from noble sires" (*Heraclitus*, 7);"If one were to yoke good with bad, no good offspring would be born, but if both parents are good, they will bear noble children" (*Meleager, Fr.* 9).

The early Romans similarly held lineage in great respect and enforced a system of *connubium*, whereby freeborn Romans could only marry into certain approved stocks. However, the Romans were relatively few in number and, when their unparalleled military and administrative ability converted the Roman empire into a fully multi-ethnic community of enormous size, the circumstances became ripe for the rise of egalitarian political ideologies. Imperial Rome, the "multicultural giant," disappeared before the onslaught of the smaller, more homogeneous, Germanic nations, which still retained a sense of group identity.

The Germanic peoples (the Germans, Dutch, Flemings, Anglo-Saxons, Franks, Lombards, Scandinavians, Goths, Burgundians and Vandals) who founded so many of the modern states of Europe following the demise of the Roman Empire, carried the concept of heredity to its logical conclusion in their virtually unique system of kinship. Unlike their kinsmen, the Greeks, Italics, Celts, Slavs, and East Balts, they did not organize themselves in patrilineal clans and phratries which recognized only their father's kinfolk, but saw kinship in fully genetic terms. The Germanic "kindred" comprised all the individual's relatives on both the paternal and the maternal sides, assessing the degree of closeness according to their actual genetic relationship; this was a quite different system from the concept of patrilineal or matrilineal clans so widespread amongst other peoples of the world. This Germanic kindred was the subject of the exhaustive study *Kindred and Clan in the Middle Ages and After* (Phillpotts, 1917). To this day most North Americans of European descent have come to accept the Germanic tradition, where kinship is determined by the closeness of genetic relationship, whether the relatives be on the maternal or paternal side, as distinct from patrilineal and matrilineal clan systems.

In ancient Scandinavia the belief in inherited talents was reflected in the concept of *hamingja,* an inherited "luck" force. However, it was recognized that siblings inherited qualities in different patterns, and kings who were "unlucky," and under whose leadership things went badly, were readily replaced by more competent individuals from the same royal lineage that had already produced generations of distinguished and successful leaders. The belief in breeding and the intergenerational transmission of genetic qualities was overriding, or as the old Germanic folk dictum expressed it, one could not make a silk purse out of a sow's ear!

Indeed, most Indo-European peoples, including those who resided outside the geographical borders of Europe, seem to have placed considerable trust in the powers of heredity. Max Weber documented the same emphasis on heredity among other Indo-Europeans. In *The Religion of India* (1958), Weber described the semi-magical *xvarenah* attributed to Indo-Iranian kings as a belief in inherited ability, calling it "familial charisma." The Indian caste system, he maintained, was sustained by a similar belief in the genetic inheritance of human qualities. The charisma of a caste, of a sib, and of a family, was genetically transmitted; its roots were to be found in the concept of inherited ability.

The coming of Christianity plunged logic and classical philosophy into centuries of near-oblivion and clashed with the established and ancient European belief in the *inequality* of men. Spreading first among the slaves and lowest classes of the Roman empire, Christianity came to teach that all men were equal in the eyes of a universal Creator God, an idea that was totally alien to older European thought which had recognized a hierarchy of competence among men – and even among the gods. Opposing the traditions of classical philosophy and scientific enquiry, Christianity introduced the concept of a single, omnipotent "God of History" who controlled all the phenomena of the universe – with men and women being creations of that God. Since all men and women were the "children of God," all were equal before their Divine Maker! Faith in the church's interpretation of supposedly prophetic revelations became more important than scientific or philosophical enquiry; and to question the church's view of reality came to be perceived as sinful.

As Max Weber expressed it, Christianity carried the anti-intellectualism of the Middle Eastern prophets to its extreme. It represented a "non-intellectual's proclamation directed to non-intellectuals."[1]

However, although the Christian intolerance under Constantine closed Plato's academy some 900 years after its establishment, on the grounds that free thought challenged revealed dogma, traditional European appreciation of the significance of heredity never completely died. Prior to Christianity, heroes and other national leaders had been regarded as superior because of the characters they acquired by descent from famed ancestors or even from the gods: most Germanic royal lineages, for example, were believed to be descended from Woden.[2] Kings and nobles were believed to inherit qualities superior to those of the average man, and to carry these qualities in their "blood." In ancient myth heroes might even challenge the gods. The Christian church, jealous of the "divinity" awarded to kings and nobles by virtue of their lineage[3] but finding it convenient to win their goodwill, avoided challenging their "divine right" but converted this into the right to rule as earthly representatives of the Christian God – for so long as they obeyed the wishes of the Church leaders as the earthly mouthpieces of that God. The "divine right" to rule with the church's approval was a very different concept from the "divine" powers that were associated with descent from a long line of proven achievers. Consequently, the idea of genetic disparity came to be subtly discouraged by the church; and the success of the church in this respect was such that under Christian tutelage those who tilled the fields began to ask the rhetorical question: When Adam delved and

---

[1] Max Weber, *Economy and Society, Volume I,* p. 631.

[2] In free, unconquered clans, tribes, and nations, kings were symbolical figures, respected rather like elder brothers, and the English word "king" derives from Anglo-Saxon *cyning,* meaning "[the essence] of the kin."

[3] To mollify the kings, who were now deprived of their claim to superiority based on the distinction of their ancestry, the Catholic Church argued that monarchs who were loyal to the Church had a "divine right to rule" on behalf of God, but only for as long as they recognized the superior moral authority of the Church.

Eve span, who was then the Gentleman?

Stripped of their belief in the significance of human heredity and the notion of the state as a kinship unit – "a family writ large" – and believing instead in the essential equality of all men and women as the children of God, dissident sects espousing radically egalitarian ideals arose at intervals to protest social and economic inequality, especially at times when this became oppressive.

In time, secular political movements also began to assert the idea of biological equality, a theme which tended to be favorably received whenever the disquietude of a divided society erupted into revolution. Such was the case of the Levellers who fought alongside the Parliamentarians in seventeenth century Britain;[4] of the Jacobins, who decimated the accomplished aristocracy of eighteenth century France; and of the Bolsheviks who wrought genocidal slaughter among the more successful members of Czarist Russian society – nobles and peasants alike – following the Bolshevik Revolution in the early twentieth century.

In recent times, calls for political revolution have frequently invoked attacks on "genetic determinism" in favor of the alternate, wildly illogical, philosophy of "biological egalitarianism." Despite the fact that both Marx and Engels personally believed in the significance of heredity and race – Marx often resorted to racist terms to abuse rivals in correspondence with his friends – the ideological movement that emerged from their teachings eventually yielded to the notion of biological egalitarianism as a necessary ploy to inspire revolutionary passions among the "proletariat." Under Stalin, who sought to spread revolution in the Third World, communist theoreticians rejected any idea that economic

---

[4] It was in the tradition of their British Puritan forebears, who placed emphasis on the spiritual equality of man under the rule of an all-powerful God of History, that the New England religious thinkers of Eighteenth Century America condemned the more hierarchical, aristocracy-favoring attitudes of the Southern plantation owners.

inequality could be explained by biological inequality. The suggestion that one individual might be inherently more creative or productive than another was unlikely to fuel the feelings of resentment necessary to incite the masses to revolutionary action.

II

THE DISCOVERY OF EVOLUTION: EUGENICS AND THE PIONEERS OF MODERN SCIENCE

YET EVEN while the myth of biological egalitarianism was gaining ground in the Western world, the momentum of scientific enquiry, freed by the Renaissance from the shackles of medieval religious dictates, was deepening Man's knowledge about himself and the world around him. In addition, a renewed enthusiasm for the application of selective breeding to plants and animals in the agricultural revolution of the eighteenth century focused enlightened thought once again on the significance of heredity.

In the second half of the nineteenth century, Charles Darwin restored the concept of heredity to its rightful place with his epic work, *The Origin of the Species by Means of Natural Selection or the Preservation of Favored Races in the Struggle for Life* (1859). His research troubled his deeply religious wife, because she sensed that it challenged the still dominant pattern of religious thought. Facing the need to defend his overall theory of evolution as applied to all living species, Darwin is described by his biographer, Sir Arthur Keith, as having decided to refrain from extending his evolutionary theory to explain the inequalities between the surviving races of man, which he regarded as being so apparent.[1]

What Darwin found it necessary to avoid, his cousin Sir Francis Galton did not hesitate to tackle. Indeed, Galton established the science of statistics as he sought to understand

---

[1] See *Darwin Revalued*, by Sir Arthur Keith. 1955. Watts & Co., London.

the process of inheritance. Galton was as great a contributor to science as was Darwin, for apart from the attention he directed to the study of heredity, he also laid the foundations for the science of meteorology, and together with his friend, co-worker and biographer Karl Pearson, he established modern statistical methods and quite literally founded the science of *eugenics.* The goal of eugenics, a word created by Galton from the Greek *eugēnes* ("well born"), was to apply scientific knowledge about heredity to the problem of human evolution in order to combat deleterious demographic trends which threatened to lead to a decline of genetic quality in modern societies. In Galton's own words, the purpose of genetic science was "to give the more suitable races or strains of blood a better chance of prevailing speedily over the less suitable." Significantly he described eugenics as "that science which deals with all influences that improve the inborn qualities of a race; also with those that develop them to the utmost advantage" (1909, 35). Eugenics, he explained, was rooted in a deep concern for posterity:

> Man is gifted with pity and other kindly feelings; he has also the power of preventing many kinds of suffering. I conceive it to fall well within his province to replace Natural Selection by other processes that are more merciful and not less effective."[2]

Using statistical techniques to demonstrate the role of genetics in shaping mankind, Galton argued that it was scientifically possible to increase the frequency of desirable qualities among human beings, and to prevent the spread of deleterious qualities, by cugenic measures, and the idea quickly attracted the favorable attention of serious scholars following the publication of his epoch-making study *Hereditary Genius: An Enquiry into Its Laws and Consequences* (1869).

---

2. Sir Francis Galton, *Memories of My Life* (New York: Dutton, 1908), p. 322.

This seminal text was followed by *Natural Inheritance* (1889) and *Essays in Eugenics* (1909). Galton also realized something that is too often forgotten by modern commentators: that nature and nurture work in tandem and are not to be seen as mutually exclusive opponents. Heredity was important, but so was a healthy and congenial environment.

It is on record that Darwin was impressed by his cousin's ideas. In a letter dated December 3, 1869, Darwin congratulated Galton on *Hereditary Genius*, describing this as a "memorable work," and observing that "I do not think I ever in my whole life read anything more interesting and original – and how well and clearly you put every point – You have made a convert." Two years later, in chapter seven of *The Descent of Man*, he echoed Galton's observations concerning the differences between human races, noting that:

> ... the various races, when carefully compared and measured, differ much from each other – as in the texture of hair, the relative proportions of all parts of the body, the capacity of the lungs, the form and capacity of the skull, and even the convolutions of the brain. But it would be an endless task to specify the numerous points of difference. The races differ also in constitution, in acclimatization and in liability to certain diseases. Their mental characteristics are likewise very distinct; chiefly as it would appear in their emotion, but partly in their intellectual faculties. Everyone who has had the opportunity of comparison, must have been struck by the contrast between taciturn, even morose aborigines of S. America and the light-hearted talkative negroes.

Thus both Darwin and Galton came to the same conclusion, expressed by Galton in *Hereditary Genius* as follows:

> It is in the most unqualified manner that I object to pretensions of natural equality. The experiences of the nursery, the school, the university, and of professional careers, are a chain of proofs to the contrary ... In

> whatever way we may test ability, we arrive at equally enormous intellectual differences.

Galton's younger colleague, Karl Pearson, developed Galton's novel statistical techniques to new levels of effectiveness, laying the foundations of modern scientific method in his publication *The Grammar of Science* (1892). Like Galton, Karl Pearson realized that the genetic legacy each generation leaves to its successors is of prime importance for the future of mankind. Every generation, in fact, is a bottle-neck which sifts and determines which genes are to survive. Pearson delineated the fundamentals of the new field of eugenic science in a number of publications, including *National Life from the Standpoint of Science* (1901), *Nature and Nurture: The Problem of the Future* (1910). He expressed his concern for the genetic future of the British nation in a warning to his fellow-Britons in his Huxley Memorial Lecture of 1903:

> the mentally better stock in the nation is not reproducing itself at the same rate as of old – the less able and the less energetic are the more fertile ... The psychical characters which are the backbone of a State in the modern struggle of nations are not so much manufactured by home and school and college; they are bred in the bone, and for the last forty years the intellectual classes of the nation, enervated by wealth or by love of pleasure, or following an erroneous standard of life, have ceased to give in due proportion the men wanted to carry on the ever-growing work of the Empire." (1903)

Early Eugenics in Britain

ANY PEOPLE who recognize the significance of heredity must naturally think in terms of breeding. Once science had revalidated the concept of heredity in the Western world, the reaction in favor of extending the principles by which the quality of plants and animals had been improved to human beings was natural. The conditions of life in modern society seemed to be reversing natural selection and lowering the quality of each succeeding human generation. Support for the eugenic ideal quickly came from a wide range of varied intellectuals, including not only traditionalists who had always retained their belief in good breeding combined with good training, but also progressive thinkers. Those who cared for the unfortunates of this world saw how simply much human suffering could be eliminated in future generations by eugenic policies, and socialists such as George Bernard Shaw, whose *Man and Superman* (1965, p. 159) (essentially an ode to the inborn instinct to procreate the race) complained of contemporary society that "being cowards, we defeat natural selection under cover of philanthropy." H. G. Wells, another reformer who likewise cared for posterity, proclaimed that "the children people bring into the world can be no more their private concern entirely, than the disease germs they disseminate" (Kevles, 92). Others who supported the eugenic ideal were the youthful J. Maynard Keynes; left-leaning Julian Huxley, who sought not revolution but the reduction of human suffering by genetic improvement; and J. B. S. Haldane, who adopted Marxist values but always opposed its anti-hereditarian extremes. Numerous other social reformers of that time, such as Sidney and Beatrice Webb, likewise embraced the eugenic ideal – they were patriotic in the tradition of William Morris and Charles Dickens and eschewed revolutionary socialism, but feared emerging capitalism as a threat to the traditional bliss of agrarian England, and felt that much misery could be eliminated by rearing fit and healthy children rather than those who were burdened by genetic handicaps.

Also joining the eugenics cause was the ardent advocate of social change, Havelock Ellis, who supported the call for female liberation but emphasized the essential role that women played in ensuring the future of the race. Ellis (1912, pps. 46–47, 205) declared that the aims of eugenics "could only be attained with the realization of the woman movement in its latest and completest phase as an enlightened culture of motherhood." The new St. Valentine, he observed, would be a scientific saint, not one of folklore, because marriage should be for the procreation and health of the race, not merely for personal pleasure. Scholars and politicians alike applauded the new sense of responsibility in procreation,[3] with diverse figures such as the Cambridge biologist Francis Maitland Balfour, founder of the British school of evolutionary biologists, British Prime Minister Arthur James Balfour,[4] and the young politician Winston Churchill, all paying homage to the eugenic ideal.

Galton, childless himself, applied his personal fortune toward the promotion of research into heredity and eugenics, funding the establishment of a biometrics laboratory at the University of London under the direction of his fellow-eugenicist Karl Pearson, for the primary purpose of studying heredity in man. He also helped finance the establishment of the Eugenics Education Society, which later changed its name to the more simple Eugenics Society. Patriotic Englishmen who feared a dysgenic trend in national ability eagerly supported the eugenic doctrine that the fittest, most intelligent and creative parents should be encouraged to have larger families. In this, they were joined by Fabian socialists, who sought to decrease what was seen to be an excessive rate of reproduction among the genetically unfortunate, so as to

---

[3] The eleventh edition of the *Encyclopedia Britannica* stated in its entry under "civilization" that future human progress would include "the organic betterment of the race through wise application of the laws of heredity."

[4] Balfour was the prime minister of Britain (1902–05) and foreign secretary (1916–19) who favored Anglo-American "Anglo-Saxon solidarity" and also made the Balfour Declaration promising a national home for the Jewish people. He delivered the inaugural address at the first International Eugenics Congress, which was held in England in 1912. Cosponsors of that Congress included Winston Churchill, Alexander Graham Bell, Charles William Eliot (a former president of Harvard University), and the Bishop of Ripon (in England). It was attended by 750 people.

"level up" society instead of "leveling it down" – which latter was the usual outcome of revolutionary socialism.

In general, the socialists and liberals favored "negative eugenics" – a reduction of the number of births among the less favored – while the traditionalists tended to think in terms of the competition between nations and races, and favored "positive eugenics," which sought to encourage a higher rate of reproduction among the better stock.

Possibly it was Julian Huxley who best summed up the confidence with which so many British academics who lived during the first half of this century viewed the future, when he wrote:

> Once the full implications of evolutionary biology are grasped, eugenics will inevitably become part of religion of the future, or whatever complex of sentiments may in future take the place of organized religion. It is not merely a sane outlet for human altruism, but is of all outlets for altruism that which is most comprehensive and of longest range (1941, p. 22)

Although it is today difficult to envisage his prophecy becoming a reality in the contemporary Western world, tendencies in Mainland China, and in the Chinese republic of Singapore, strongly indicate that it may be the billion-plus Chinese people who first realize Huxley's dream of the future.

The Eugenic Ideal Finds Favor in America

Scientific ideas are seldom long confined to one country in the modern world. Although it was in England that the concepts of evolution and eugenics first saw light, European and American scholars soon responded. We will not here attempt to cover the continental scene, although scholars such as Ernst Haeckel, who became an ardent advocate of Darwinian evolution, seeing nations as potentially incipient races and the major racial divisions of mankind as virtually separate species, undoubtedly influenced the English-speaking world. At this time the determination of what constituted a species had not yet come to be linked to the concept of mutual inter-fertility, but was judged purely by the extent of phenotypical variation, as in the Linnaean system of classification still broadly accepted by biologists today. Consistent with such views, Haeckel and others began to urge not only eugenic breeding but also racial purity.[5]

The concept of a new eugenic science was also welcomed in the United States, which shared the same traditional appreciation of the role of heredity held by those Europeans who had remained behind in Europe. At the turn of the century, the United States was a land of opportunity, one which had already acquired a sense of nationhood, so that many of its most important families had developed a profound social conscience and a strong desire to ensure that the hopes they held for the well-being of their descendants, as Americans, would be realized. Idealists such as president Theodore Roosevelt were convinced that the existing American population possessed generally superior genetic qualities, shaped by severe selective evolution over the previous generations. Their forebears had been adventurous individuals who had first elected to undertake, and then survived, what was in earlier centuries a dangerous ocean crossing. After arrival in the New World, they had to protect themselves and their families from the depredations of the native Indian tribes who had the advantage of familiarity with the local environment. While doing this, they had to tame vast primeval forests and grassland wildernesses – something Europeans had not seen since their forebears first began to convert the forests of Europe into the rich but increasingly overpopulated farmlands of civilized agrarian and mercantile culture. Thus a century and a half ago, Ralph Waldo Emerson reflected the views of his countrymen when he wrote that: "If

---

[5] Various of Haeckel's works have been translated into English, notably The Evolution of Man, 1905

the race is good, so is the place".[6]

Americans at that time did not think of their country as a potential microcosm of all humanity, but as an emerging micro-race of predominantly European origin. President Theodore Roosevelt, credited with advancing the "melting pot" ideology, wanted only quality immigrants from ethnic stocks which would readily assimilate into the "Old American" population – a term used to refer to persons descended from Europeans who had settled in North America prior to the War Between the States. While the British eugenicists were primarily concerned with maintaining the breeding quality of the resident population of the British Isles, Americans also debated the question of immigration, since they instinctively knew that immigrants affect what we would today call the national gene pool quite as significantly as differential selection within that pool.

Like Theodore Roosevelt, eugenicists felt that the new America must remain a vital and homogeneous nation. But, Roosevelt strongly believed that the Old Americans were not producing enough children, and that they must either change their ways or submit to an invasion of non-white peoples, most likely from Asia. Selected immigration from Europe was welcome, but those who would not fit in were not wanted. Immigrants should desirably match the genetic character of the existing population, and, to ensure this, he favored the restriction of immigration to the nations from which the predominantly North European pioneers who had built the United States had been drawn. The eugenic ideal matched perfectly the optimistic, forward-looking spirit of the people of the United States as they entered the twentieth century (although Roosevelt was fearful that those who advocated negative eugenics might discourage large families).

When eugenicists looked at the increasing level of Asian and Hispanic immigration already then apparent, some feared

---

[6] The Complete Writings of Ralph Waldo Emerson. Wm. Wise & Co. New York. 1929. p.441

that the "great race" – as eugenicist and conservationist Madison Grant[7] described those whose ancestors had pioneered the establishment of European civilization in North America – might be swamped by the new immigrants of diverse national origin and different genetic stock, too numerous to be assimilable.

Madison Grant's own ancestors, it might be noted, had come to the American colonies from Scotland following the failure of the 1745 Highland uprising led by Bonnie Prince Charlie. His writings were therefore well received by a generation of proud, self-confident, and essentially prosperous Old Americans who wished to see their country remain in the hands of their own kind, and who like the Greeks of old treasured the memory of the achievements of their forebears. American scholars, wealthy self-made industrialists, farmers, and even politicians saw the eugenic ideal as a means of ensuring the future well-being and happiness of the new nation to which they were proud to belong. Indeed, it was those who could claim to be Old Americans who gathered most enthusiastically in support of the eugenic cause.

The Institutionalization of Eugenic Ideology

THE HOPES OF THE eugenicists were raised in 1910 by the establishment of the Cold Spring Harbor Eugenics Record Office by the Carnegie Institute of Washington. This was funded by Mrs. Mary Harriman,[8] the widow of E. H. Harriman, whose forebears left England for America in the seventeenth century. The director was Charles B. Davenport,

---

[7] Grant was well known by literate Americans for his books on racial history, such as *The Passing of the Great Race or the Racial Basis of European History* (1918) and *The Conquest of a Continent* (1921).

[8] In 1910 Mrs. E. H. Harriman established the Eugenics Record Office at Cold Spring Harbor, purchased a farm of about 80 acres with buildings near the Station for Experimental Evolution, and in 1913 paid for the erection of an office building on it. On January 1, 1918, she transferred the office and farm to the Carnegie Institution and established an endowment fund of $300,000 toward its maintenance.

the Harvard zoologist who was twice president of the American Association for the Advancement of Science (AAAS), as well as president of the American Zoological Society. The superintendent was Harry H. Laughlin, a leading light in the eugenics movement which flourished in America during the first half of the present century.

The distinguished inventor, Alexander Graham Bell, figured prominently among the members of the Board of Scientific Directors established to support the work of the Eugenics Record Office. In a letter to Davenport, dated December 27, 1912, Bell revealed himself as a "mainstream" eugenicist who believed in "positive eugenics," which aimed at increasing the percentage of healthy and talented individuals in succeeding generations, rather than in "negative eugenics," the term commonly ascribed to measures designed to prevent the spread of deleterious genes.[9] In light of the somewhat limited development of genetic and medical science at that time, and the heavily dysgenic impact of two World Wars which were shortly to follow, his observations reflect the perspicacity of a scientist whose name will live forever in history as a major contributor to technological progress and as a benefactor of the human race.

Bell attended eugenics conventions, and himself authored several papers on eugenics, such as his essay entitled "How to Improve the Race" (1914) in *Journal of Heredity,* edited by Paul Popenoe.

The Galton Society – formed in New York in 1918 at the American Museum of Natural History by Henry Fairfield Osborn, C. B. Davenport, and Madison Grant – became actively involved in endeavors to shape U.S. policy relating to population quality. Spontaneous eugenics societies were established in many American cities, and hopes for a bright future for eugenics and future generations were raised when in 1923 (due largely to the efforts of Alexander Graham Bell, Luther Burbank and Charles B. Davenport) the American

---

[9] A copy of this letter is in the possession of the author.

Eugenics Society was established, with branches in numerous American cities. With the foundation of the Society, the eugenics movement began to take shape in a businesslike manner, placing heavy emphasis on the participation of scholars as scientific advisors on various advisory committees. Other eugenics organizations proliferated at this time, among which were the Eugenics Research Association, the International Federation of Eugenic Organizations, the American Social Hygiene Association, and even such bodies as the American Genetics Association, and the American Association for the Study of Human Heredity. Eugenicists were kept informed of new developments in science through a publication entitled the *Eugenical News*.

Nations as Breeding Populations

DEVOTED to the well-being of their young nation, many of the early American eugenicists were proud to claim to be of Old American stock.[10] The term "Old American" was further popularized by the book of the same name, compiled by the anthropologist Aleš Hrdlička. Hrdlička was himself a recent European immigrant, of Bohemian origin, who had accompanied his father to America at the age of fourteen and, after working in a cigar factory as a teenager, had entered the Eclectic Medical College of New York and graduated with an M.D. at the top of his class. Hrdlička practiced medicine for a while but soon found his interest turning to the problem of human quality and the significance of racial differences, which he saw in an evolutionary context.

Traveling extensively to study the diverse living peoples

---

[10] Aleš Hrdlička's *Old Americans* (1925) comprised a detailed anthropometric study, accompanied by photographs, of individuals who were in general descended wholly from not less than three generations of persons born in the United States, thus representing the descendants of settlers who had arrived in the North American continent prior to the War Between the States, and in many cases prior to the War of Independence.

of the world, as well as paleontological remains, he emerged as America's leading physical anthropologist, serving as editor of *The American Journal of Physical Anthropology* (which he founded in 1918), and as first president of the American Association of Physical Anthropologists. Significantly he found nothing wrong in the historical pride of his new compatriots, and served for many years on the American Eugenics Society subcommittee on anthropometry and on the advisory council, being deeply concerned, as an expert on evolution, with the threat of dysgenics (of a deterioration in genetic stock) facing modern man.

In his article "Race Deterioration and Destruction with Special Reference to the American People," published by the Race Betterment Foundation of Battle Creek, Michigan as part of the proceedings of the Third Race Betterment Conference held on January 2–6, 1928, Hrdlička defined "race deterioration" as "the degradation of its standards of mentality and effectiveness, generally attended also by the lowering of those of physique" (p. 82). "Race destruction," another threat with which he was concerned, meant the loss of racial identity either "through complete submergence into another race" (p. 82) or simply by failure to reproduce itself.

History, Hrdlička noted, recorded the rise and fall of nations. Vital, healthy nations tended to rise, but nations could also fall when, weakened by luxury and the exhaustion of ambition, they suffered a "dilution of the physical as well as mental status by admixture with poorer blood" (p. 83). He observed (pp. 84–85):

> There are still some benevolent minds who would like to see all men, white and black, potentially equal. Yet they will hold that there are differences between one family and another family, and even between the children of the same family, in the same racial group. If they did not, there would obviously be no use for eugenics ... no use even for much of genetics and biology. As a matter of fact there are similarities but no absolute equality anywhere in living nature, either in races, or families or

> even individuals. The problem is merely how great in a given case is the dissimilarity. Races, especially the further distant ones ... are not equipotential ...

The Race Betterment Foundation, which sponsored the conference at which Hrdlička presented these views, was first established in 1906 as the American Medical Missionary Board (but soon changed its name to the Race Betterment Foundation). Its founder was John Harvey Kellogg, a descendant of an Englishman named Joseph Kellogg who had arrived in North America as early as 1651. Kellogg, who launched the breakfast cereals industry by introducing granola to the American public as a health food, was chief surgeon at the then world-famous Battle Creek Sanitorium.

Publishing a journal called *Good Health,* the Race Betterment Foundation became a major center of the new eugenics movement in America. Kellogg himself was an important and respected figure who authored numerous medical and eugenics treatises, and his circle of influence extended to several successful businessmen including J. C. Penney and C. W. Barron, whose names remain familiar to this day. Another member of the Kellogg family, Vernon Lyman Kellogg, a zoologist of international repute, also espoused the eugenics movement. As a personal friend of President Herbert Hoover, he served on various national health and agricultural committees, becoming a trustee of the young Brookings Institute and of the Rockefeller Foundation while continuously taking an active role in the eugenics movement as the latter grew in size and influence.

The movement was also early supported by famed educator David Starr Jordan, first president of Stanford University.[11] Jordan's first American ancestor had arrived in

---

[11] Jordan realized, just as World War I was breaking out in Europe, that modern warfare exercised a severely dysgenic influence on participant nations, and World War I certainly proved him correct. He

the North American colonies from England circa 1700. Jordan's status in the American education scene of his day is illustrated by the fact that he was a trustee of the Carnegie Foundation and president of the American Association for the Advancement of Science, the National Educational Association, and the California and Indiana Academies of Science, as well as vice president of the Eugenics Education Society in Britain. Charles William Eliot, president of Harvard University, trustee of the Carnegie Institution of New York, and one of the most eminent educators in America, also rallied to the eugenic crusade, as did astronomer William Wallace Campbell, president of the University of California, whose Scottish forebears had migrated to the colonies in the eighteenth century. Livingston Farrand, president of the University of Colorado and subsequently president of Cornell University, chairman of the central committee of the International Red Cross, and editor of the *American Journal of Public Health,* similarly espoused the eugenics movement, as did innumerable other educators and faculty members of note.

Realistic Anthropology

AMERICAN PALEONTOLOGISTS and anthropologists recognized the significance of evolution – as well as the possibility of devolution – and were generally enthusiastic about the idea of eugenic policies. Another leading anthropologist who served on the American Eugenics Society subcommittee on anthropometry was Earnest A. Hooton, the respected Harvard professor who later became director of the

---

wrote several books on the subject, his *War and the Breed* (reprinted in 1984 and available from Cliveden Press, Washington D.C.) being completed in 1915, the second year of World War I when young men, the cream of European youth, were dying by the millions in France and Belgium – most too young to have married and left offspring. Britain, in particular, had a purely volunteer army for the first two years of the war.

Peabody Museum. Hooton's father had been born in England, and Hooton himself attended Oxford University on a Rhodes scholarship. Returning to America, where he became one of the main pioneers of physical anthropology, Hooton was an active member of the American Genetics Association and the Galton Society. He saw eugenics in terms of different levels of human evolution, and strongly believed that different inherited personalities contributed to susceptibility to engage in criminal behavior. Hooton collated data on some seventeen thousand criminals, and took a keen interest in the effects of race-mixing. His *Up from the Ape* (1931) and *Apes, Men and Morons* (1937) argued that heredity was at least as important as environment as a determinant of human behavior.

Most scholars who were in any way connected with the study of evolution came to see races as representing different levels of human evolutionary development. Ellsworth Huntington, a Yale geologist descended from one, Simon Huntington, who had emigrated from England to the colonies in 1633, was at one time or another president of the Association of American Geographers, president of the Ecological Society of America, director of the Population Association of America, member of the American Eugenics Society advisory council and chairman of its Committee on Biologic Genealogy.[12] A widely traveled scholar, Huntington expressed his fear that the more advanced human stocks would likely be overrun by the less advanced, and, as a prolific author (he wrote no less than twenty books and several hundred articles), his views reached a broad audience of educated men and women.

Biologists were naturally prominent among those who

---

[12] Huntington became deeply immersed in the eugenics movement and in his book *Tomorrow's Children* (1935, 72–73), he observed that "[i]t is hard to see how a perfect eugenic system can prevail until every intelligent married couple is able to have as many children as it wishes without lowering its economic status."

recognized the role of heredity in human affairs, and many adopted the eugenics cause. Notable amongst these was H. S. Jennings of The Johns Hopkins University, himself of Old American stock, who wrote several influential books on the subject: notably *Prometheus or Biology and the Advancement of Man* (1925), which forecast a bright future for mankind through the application of eugenic policies, and *The Biological Basis of Human Nature* (1930). Jennings was president of the American Society of Zoologists and the American Society of Naturalists, as well as a member of the advisory committee of the American Eugenics Society.

Psychologists Welcome Eugenics

PSYCHOLOGISTS who were interested in intelligence also tended to become involved with the new ideal of population improvement. R. M. Yerkes, a descendant of Anthony Yerkes who had come to America from Holland in 1700, became an enthusiastic member of the American Eugenics Society and the Galton Society. A professor of psychology at Harvard, Yerkes is particularly known for his mammoth World War I study of the IQ ratings of one and three-quarter million U.S. military recruits, compiled while he was head of the psychology division of the office of the United States Surgeon General. Yerkes's concern with the different qualities of individuals and races led him to become active in the matter of immigration control, and he was elected chairman of the Committee on Human Migration in 1922. Leaving Washington to take a professorship at Yale, Yerkes recognized the potential for sociobiological studies in helping to explain human behavior, and in 1929 he founded the Laboratory of Primate Biology in Florida for this purpose. He also served as chairman of both the American Psychological Association (1916) and the American Society of Naturalists (1938). Carl C. Brigham, professor of psychology at Princeton University and author of *A Study of American Intelligence* (1923) was another member of the American Eugenics Society who became involved in immigration control, seeking to ensure that the American gene pool would not be adulterated by inferior genes, and was a keen member of the Galton Society and the Eugenics Research Association.

Possibly the most influential of the early psychologists who were active in the eugenics movement was William McDougall. Born in England, and educated at the universities of Cambridge and Göttingen, McDougall taught at Oxford University before eventually emigrating to the U.S. to take up a position at Harvard. With his experience of anthropological work in Borneo, McDougall was a fellow of the prestigious Royal Society in Britain and the author of numerous major textbooks which earned him preeminence in the field of social psychology. McDougall's respect for heredity showed itself in a series of books, beginning with *An Introduction to Social Psychology* (1908). Acutely conscious of the role of heredity in shaping human behavior, McDougall was an evolutionist who realized that human behavior was shaped by the evolutionary past of the human race. Because of the significant degree of racial diversity within the living peoples of the world, he prophesied that "racial psychology" would one day become a recognized field of study. As a believer in the quality of the North European stock relative to diverse other populations in the contemporary world, McDougall also took an interest in the composition and quality of the future population of America: his book *Is America Safe for Democracy?* (1921) stressed the influence of genetic qualities on the destinies of nations, and the need to take race into consideration when attempting to understand history or plan national policy on immigration.

Of equal significance, however, was Lewis Madison Terman, president of the American Psychological Association and of the National Academy of Science. Terman authored a number of popular books on psychology, sex, and mental health, but is academically best known for his revision of the Binet Scale (1916), his co-authorship of the Stanford Achievement Tests, and his massive four-volume *Genetic Studies of Genius* (1926–30). Terman was a major voice in the eugenics movement, and was a key member of the Eugenics Society committee on psychometry.

But possibly it was Charles Conant Josey of Dartmouth College, who in his capacity of professor of psychology most strongly emphasized the importance of heredity and race in determining the fate of nations and civilizations. His 1923 publication, *Race and National Solidarity,* stressed the significance of heredity and also of nations as cohesive micro-races in determining the trend of history. Concerned about the declining fertility rates in the Western world, and the aleady apparent signs of a future population explosion in what in recent decades has come to be known as the Third World, Josey became a bitter opponent of internationalism and the trend toward lowering national barriers to permit mass movements of surplus population around the world.[13]

Sociologists Also Note the Significance of Race and Heredity

NOTABLE SOCIOLOGISTS also rallied to the logic of eugenics. Franklin H. Giddings, author of several major works in early American sociology, professor of sociology at Bryn Mawr College, later chairman of the sociology department at Columbia University and president of the American Sociological Society, was a strong supporter of the eugenics movement who helped organize some of the first international conferences on eugenics and population. Giddings was descended from George Giddings who came to the colonies from England as early as 1635. Frank H. Hankins, a Columbia University educated sociologist who taught at Clark and Smith Colleges, was an active member of the board of directors of the American Eugenics Society.

---

[13] Josey believed in the high level of heritability of personality, and in 1922 also published *The Social Philosophy of Instinct,* which was very much in line with McIver's thinking. His Race and National Solidarity was subsequently reprinted under the title of *The Philosophy of Nationalism* by Scott-Townsend Publishers, Washington D.C., in 1995.

President of the American Sociological Society, and of the later-formed Population Association of America, Hankins authored a penetrating study entitled *The Racial Basis of Civilization* (1926).

Edward Allsworth Ross, a major figure in the development of American sociology, and author of many classroom texts, such as *The Outlines of Sociology* (1923), was deeply concerned with biology and especially population. Amongst his many publications, he wrote *Standing Room Only!* (1924), which accurately forecast the population explosion which is today seen by many as a potential threat to the ecological basis of all higher life on this planet.

Another sociologist who firmly believed in the importance of heredity was Robert M. MacIver, a Scottish immigrant who had been born in Stornaway on the isle of Harris. Well known for his many sociology texts, MacIver taught at Barnard College and Columbia University, and later was president of the New School for Social Research. A dedicated humanist, MacIver recognized the importance of good heredity and enthusiastically served on the board of the American Eugenics Society. In his major text, *Society* (1937) and other works he maintained that although material products were readily diffused from one people to another, *cultures* tended to be specific to particular populations in that they were linked to temperament and personality.

The renowned sociologist and social psychologist Emory Stephen Bogardus likewise served as an enthusiastic member of the American Eugenics Society advisory council. A professor at the University of Southern California, Bogardus edited the *Journal of Sociology and Social Research* for over forty years, founded and edited the *Journal of Applied Sociology,* and served as a contributing editor to the *Journal of Social Forces* and the *Journal of Educational Psychology.* He authored numerous textbooks which were widely used throughout the English-speaking world, and his *Development of Social Thought* (1960), written toward the end of his life, remained a classic survey of the history of sociology right into the fourth quarter of the present century. Less remembered today is his *Immigration and Race Attitudes* (1928), which more clearly shows his personal convictions concerning the importance of an understanding of heredity to any rational determination of national policy.

III
EUGENICISTS AS CONSERVATIONISTS

TO TRULY APPRECIATE the social conscience of those who supported eugenics, it is revealing to note that those who desired to preserve the genetic quality of the human population were also anxious to preserve all that was wonderful in the world of nature. Indeed, it was mainly those who were anxious to conserve the genetic heritage of mankind who pioneered the conservationist movement – which similarly sought to conserve the rich variety of plant and animal species that nature had bequeathed to the care of man.

Possibly the most notable figure in this conservationist movement was Gifford Pinchot – a grandson of one of Napoleon's generals, who was a professor of forestry at Yale and who had the honor of having Mount Pinchot named after him. A respected and popular figure, who was twice elected governor of Pennsylvania, Pinchot combined his services to the environment with active engagement in the eugenics movement. The list of eugenicists who were active in the conservation movement is a long one. The foundation of the American Bison Society was largely due to eugenicist Madison Grant and his like-minded personal friends. The Save-the-Redwoods League owed its existence to eugenicists Madison Grant, Henry Fairfield Osborn, and paleontologist John Campbell Merriam (president of the American Paleontological Society and of the Geological Society). Merriam, who was proud of his father's "Old American" ancestry and his mother's Scottish ancestry, also served as president of the Carnegie Foundation of Washington, and as an extremely active president of the Save-the-Redwoods League. Despite a full calendar, he seldom missed a meeting of the Galton Society or of the Eugenics Advisory Council.

With the support of Theodore Roosevelt, eugenicists Madison Grant and Henry Fairfield Osborn founded the New York Zoological Society in 1893, where animals could roam freely in conditions similar to their natural habitats instead of being confined to cramped cages. Irving Fisher of Yale – president of the Eugenics Research Association in the 1920s – was a member of Theodore Roosevelt's National Conservation Commission in 1919. Antioch College president Arthur Ernest Morgan, a civil engineer by profession, was another keen eugenicist who supported the conservation movement. Botanist and eugenicist James Arthur Harris, of "Old American" English stock, a student of Karl Pearson in London and winner of the Weldon Medal (named after the noted English eugenicist), was also an ardent conservationist, and served as president of the American Society of Naturalists. Perhaps even more typical of the academic naturalist and eugenicist of the day was Francis B. Sumner, a sociologist who began teaching at City College, New York, and became director of the laboratory of the U.S. Bureau of Fisheries, before moving to the University of California, La Jolla. Sumner combined his various duties as a section chairman of the AAAS and president of the Western Society of Naturalists with active membership of the American Genetics Association and the Save-the-Redwoods League.

It is only logical that eugenicists should be ardent conservationists: the humanitarian idealism of the first decades of the twentieth century was the natural outcome of discerning and cultured men and women. These were people who deemed it their duty to leave the world a better place than they had found it – or if that were not possible, at least as good a place as previous generations had left it. Many eugenicists were conservationists first and last, and there were few eugenicists who did not care about the survival of a rich legacy of flora and fauna in a healthy environment as a logical extension of their concern that the earth should be in the care of persons capable of appreciating its qualities and conserving these for future generations. Much of what has survived to the present day is a result of their efforts.

The success of the conservationist movement was facilitated by the sympathy of President Theodore Roosevelt who was deeply concerned about the threat to the quality of both the natural and human stock of America. The contemporary generation of "Old American" pioneering origin, he held, were guilty of committing race suicide by allowing their birth rate to drop below the rate necessary to populate the lands their forebears had won.[1] But he also cared deeply for the zoological and botanical heritage, and inviting George Bird Grinnell, a pioneer conservationist who had been publicizing the need for conservationist action through the medium of *Forest and Stream* magazine, to dinner to meet like-minded individuals.

Roosevelt formed the highly respected Boone and Crockett Club, without the activities of whose members, such as Madison Grant and Gifford Pinchot, there undoubtedly would be no Yellowstone Park today. Nor would the early conservation laws which saved much wild life, trees, plants, and wildernesses from destruction have been enacted. Glacier National Park is another example of the treasures that exist mainly due to members of the Boone and Crockett Club.

With Madison Grant serving as secretary and later as president, the Boone and Crockett Club was largely comprised of eugenicists and eugenics sympathizers. Renowned as one of the more active members of the eugenics movement, and especially for his efforts to preserve the "Old American"

---

[1] The term "race suicide" was invented by the noted sociologist and eugenicist Edward Alsworth Ross, author of *The Outlines of Sociology* (1923), before that discipline fell under the domination of antihereditarians, and president of the American Sociological Society between 1914 and 1915. Ross was an active eugenicist deeply concerned with population problems. In addition to serving on the Advisory Council of the American Eugenics Society for many years, he wrote two important books, *The Old World and the New* (1914) and *Standing Room Only!* (1927). The population trends now taking place in the Third World underscore the perceptive foresight of this work by Ross.

component of the American population, Grant worked just as ardently to preserve the natural heritage for future generations of Americans.

As to legislation, the members of the Club worked both inside the Club and outside to ensure the passage of federal and state laws that would protect a wide range of animal and plant life – and notably the giant redwoods – from extinction. Their fear was that the increasing population of North America and increasing economic development would destroy the heritage nature had bequeathed to their generation. They also realized that this could happen worldwide. As Madison Grant prophetically warned in *Trail and Camp Fire* (1904), unless active steps were taken:

> [i]t may be confidently asserted that [man] ... will have destroyed most, if not all the large African fauna – certainly including the most beautiful antelopes in the world – and game in India and North America in a wild state will almost have ceased to exist.

Henry Ford was another creative American who showed concern not only for the quality of the human population but also for the environment. He assigned one of his staff to act as a full-time lobbyist to support a major conservation bill which was signed into law by President Taft on March 4, 1913. Similarly it was not coincidence that Henry Fairfield Osborn, one of the most active exponents of eugenics, was among those who worked most strongly for passage of the bill.

The opening decades of the Twentieth Century were in many ways a credit to the flowering of American conservationist idealism. Led by patriotic and far-seeing descendants of the early European settlers of North America, and with the whole-hearted collaboration of the nobler-minded among more recent European immigrants, there was a strong dedication to the need to hand down a rich heritage to future generations of Americans. To have personally known any of these farsighted American leaders from the early part of this century is a privilege now denied to all except a few more elderly citizens. The names mentioned are but a few of the dedicated individuals who graced early twentieth century American society and who freely gave their time, talents and money to the effort to save for posterity all that was best of the plant, animal, and human wealth still plentiful in their day. These men and women were inspired by a noble sense of public duty, and although subsequent generations have not preserved the high level of genetic and environmental conservation to which they aspired something at least of their spirit still exists, and the overall logic behind their views has since been demonstrated to have been totally sound.

IV

RACE AS A NATION-BUILDING IDEAL

MANY EARLY SOCIALISTS, such as William Morris in England, envisaged nations as they had originally been – as kinship units whose members ought properly to care for each other and for the children who represented the nation's future. Both the men and the women thought this way, and the first women who in the nineteenth and early twentieth century spoke out for their fellow women thought the same way: they were interested in the racial future of the nation, not in internationalism. Enlightened males respected women as the progenitors of the race, and although there was a widespread concern that encouraging women to join the "work force" might reduce the birthrate, and so promote the possibility of eventual racial suicide, women activists made up an appropriately high proportion of those who attended eugenics lectures in both America and England, expressing their concern for the future of the children they bore and the nation they nourished.

Indeed, feminists such as Margaret Sanger, who for eugenic reasons wanted to make contraceptives equally available to the poor as to the middle and upper income groups, agreed with eugenicist Margaret Stopes when the latter proclaimed: "[m]ore children from the fit, less from the unfit – that is the chief issue of birth control".[1] Traditionalists such as the British husband and wife team, W. C. D. and C. D. Whetham, authors of *The Family and the Nation: A Study in Natural Inheritance and Family Responsibility* (1909), nevertheless feared that feminism and

---

[1] Quoted from Hall, 1977.

the entry of the more intelligent women into the professions would reduce the birth rate among those women who should be having more children: "Woe to the nation whose best women refuse their natural and most glorious burden," they wrote, warning that "freedom from marriage and reproduction ... is suicidal" (pp. 198–99).

The British paleontologist, Sir Arthur Keith, particularly saw nations as separate breeding populations, each potentially emerging as mini-subspecies, and warned against any dysgenic decline in intelligence. Indeed, he suggested that dysgenic trends may have been responsible for the collapse of the urban civilizations of antiquity:

> Babylonia, Egypt, Greece and Rome did not fail for lack of children, but for a dearth of children of the right kind. There is no doubt that civilization exposes the ablest of her votaries to an extreme temptation – the temptation of luxury. The basal and essential condition for racial progress is a healthy birth rate – in every grade of every community. As we raise our standard of living, the more expensive becomes the maintenance and education of our children; we become more and more tempted to spend on ourselves what should be sacrificed in raising another generation.[2]

The threat to civilized nations of a possible biological degeneration was echoed by the Nobel Prize-winning geneticist, Hermann J. Muller:

> Another way in which lack of a very highly developed social impulse must in the end lead to the downfall of any civilization so afflicted lies in its biological undermining.... there are indications that under civilization many traits that are detrimental to the

---

[2] Sir Arthur Keith, *Concerning Man's Origin*, 1828, London: Putnam's Sons. p. 125. See also his *Evolution and Ethics*, 1947, New York: Putnam.

> species as a whole are propagated at a *faster* rate than the normal, so that the otherwise exceedingly slow biological degeneration is thereby hastened. This, when the increasing complexities of living are really calling for even greater ability on the part of the group as a whole.[3]

In Britain, the distiguished research psychologist Raymond B. Cattell provided convincing statistical evidence of the threat to Britain of declining intelligence in *The Fight for our National Intelligence* (1937),[4] and ranked occupations by the mean intelligence of those engaged therein, in a study published in *Eugenics Review* in the same year.[5] This identified the range of intelligence of workers in diverse occupations, showing that those in the intellectually more challenging professions (who generally produced only small numbers of children) exhibited higher levels of intelligence than artisans and laborers. William McDougall, the noted psychologist, also expressed concern about the fact that in complex societies the more intelligent generally failed to reproduce at even replacement level:

> Looking at the course of history widely, we may see in the tendency of the upper strata to fail to reproduce themselves an explanation of the cyclic course of civilization.[6]

The Threat of Declining Birthrates

TODAY, the low birth rate among the Western and

---

[3] H. J. Muller, *Out of the Night*, 1935, New York: Vanguard, pp. 42-43.

[4] R. B. Cattell, *The Fight for our National Intelligence*, London: King & Co., 1937.

[5] R. B. Cattell "Some Relations of Intelligence to Socioeconomic Factors", *Eugenics Review*, 1937.

[6] William McDougall, *Group Mind* (London: Cambridge, 1939), p. 260.

more advanced nations of the world, and the massive ongoing population explosion in the Third World, have rendered the views of these early eugenicists prophetic. Even the Chinese of Singapore, who have attained a high level of civilization and sophistication have noted the low birth rate among the more intelligent and educated of their womenfolk. As Leonard Darwin, Charles Darwin's son, a leading supporter of the British Eugenics Education Society and the British *Eugenics Review*, warned in 1927, the spread of birth control devices has been "racially devastating" to the more advanced countries. Sir Charles Galton Darwin, grandson of the renowned evolutionist and one of the early editorial board members of *The Mankind Quarterly*, also stressed the threat to civilization of the vast increase in population in the Third World as a result of modern medical advances in his book *The Next Million Years* (1951) and in his 1960 *Mankind Quarterly* article "World Population: Can Man Control His Numbers?"

In the early decades of this century, both in America and in Britain, there was a generally happy overlap of interest on the subject of eugenics between leading liberals, who espoused negative eugenics aimed at discouraging the genetically defective from procreating, and traditionalists such as Leonard Darwin and Coldstream Guards officer C. P. Blacker, who supported positive eugenics and who were more concerned that the talented and healthy should pass on their genes to future generations. This political overlap was matched by a softening of the attitude of the Church, with the Bishop of Ripon urging the greater procreation of the fit in order to populate the British dominions overseas. The same moderation of attitude toward eugenics among the more progressive Christian churchmen of this time was also to be found in America – a notable modification of the Christian tradition which was not adopted into present-day Christian ethics. As Rudolph M. Binder wrote in his 1915 article "Eugenics and Religion," which was published in the *Eugenical News*:

> There has always been a double line in theological reasoning. One holds that God is not only omnipotent but controls absolutely everything; and all the various phenomena in the universe, including human activities are but manifestations of his power. This theory is best known by the term "predestination," and would imply that ... any attempt at contraception is an interference with the will of God. Hence, the opposition to eugenics which, while primarily concerned with the increase of good stock, is at least indirectly opposed to the propagation of poor stock.
>
> This theological theory has gradually yielded to a more ethical conception of the deity. The omnipotence of God is less emphasized than His love. And the new theory permits a different interpretation of man. He has some liberty, he is held responsible for his acts, he is praised for his good actions as a co-worker with God and blamed for his bad ones.

With various other religious leaders such as Dean Inge of St. Paul's Cathedral supporting eugenics, the years between 1900 and the early 1930s were generally marked by a happy collaboration among those who desired to improve the conditions of life for future generations.

Those were days, it must be remembered, when concern about the "yellow peril" was common. This was based on the evidence, already apparent but since then magnified many times, that while the white race was threatened by a decline in numbers and quality, the speed with which the population of Asia was increasing, and the attempts Asians were already making to migrate into North America, constituted a threat to the United States which then perceived itself as a white nation. G. Stanley Hall, president of Clark University, was among those who warned that the traditional character of America as a European-based nation was threatened by the uncontrolled immigration of Asians – logically and prophetically commenting that "the future belongs to those people who bear the most and best children and bring them to maturity". In Britain the Whethams similarly warned against the immigration of genetically quite diverse peoples[7] and echoed the same theme: that the future belongs to the more prolific, and that, if the disparity in births continued, the world could one day be dominated by the Chinese.

The More Advanced Countries Adopt Eugenic Laws

In the early decades of the twentieth century, eugenicists knew that if the members of their generation were to pass on their genetic heritage to posterity in at least as good a condition as they had received it, it was necessary to avoid an undue proliferation of deleterious genes and to ensure the transmission of the more desirable genes. It was obvious to the more conscientious that modern conditions of life had undermined nature's own methods of preserving the quality of the human stock in those populations that had emerged from a feral state of life and had advanced into what they so proudly called a "civilized" existence. Something had to be done to block the dysgenic trends that were weakening the quality of the population in the more technologically advanced nations. Remote tribes living in the Amazon were still subject to nature's pruning knife and did not yet face these dysgenic trends, but they too, as their lifestyle became "modernized," were likely in the course of time to face dysgenic influences.

As a result, the more responsible members of the advanced Western nations, conscious of their duty to the well-

---

[7] Interestingly, America and many other countries are today experiencing increasing illegal Chinese immigration, with some sources estimating thirty thousand *illegal* Chinese immigrants entering the U.S. every month. The British government recently promised to accept an estimated one-third to one-half million Chinese from Hong Kong when that territory is surrendered to Mainland China. Mainland China, not counting the Chinese on Taiwan or those already settled in large numbers in other Oriental countries, is now on its way to attaining a population of one and a quarter billion, and could alone fill all the empty quarters of the world without the assistance of any other peoples!

being of future generations, came to press for eugenic laws, and the majority of the more enlightened nations of the West introduced some such kind of legislation. Some might argue today that they should have waited until genetic science had advanced to the level where genetic "surgery" had become a real possibility, but the immediacy of the dysgenic threat introduced by modern conditions did not permit the luxury of delay. The entire genetic potential of future generations was going to be restricted to whatever collation of genes those who were living in the early decades of the twentieth century passed on to their heirs. Eugenicists realized that the future genotype of a nation depends on the reproductive activity of each successive generation, and the massive dysgenic impact of modern warfare, combined with the emergence of disproportionate rates of reproduction between the fit and the unfit, indicated that there was no time to be lost.

In consequence, most nations of the Western world introduced eugenic laws to control the spread of deleterious genes. Eugenic laws were introduced in Denmark, Norway, Finland, Sweden, Lithuania, Estonia, Iceland and a number of Swiss cantons. Germany introduced similar laws and has been much criticized for this, but so also did Austria, Hungary, Italy, Greece, and Spain. Britain, which had largely pioneered the eugenics movement, did not get around to following suit, possibly because it had already lost a large part of its leadership class in World War I. Outside Europe, there were countries such as Argentina, Chile, Peru, Brazil where laws covering hereditary mental pathologies, hereditary feeblemindedness and serious hereditary physical pathologies were enacted.

It is of some interest that these legislative measures, designed to protect future generations, were largely pioneered by the United States; U.S. legislation being copied by many of the other countries, including Germany. By 1931 twenty-seven of the forty-eight states of the U.S. had eugenics laws on their statute books, and no less than thirty states in all passed such laws at one time or another. In addition, many states enacted statutes prohibiting miscegenation between blacks and whites, based largely on the belief that the scrambling together of disparate gene pools created by nature would be dysgenic. In Canada, both British Columbia and Alberta adopted eugenic sterilization, and it seemed probable that in the course of time all the leading nations of the West would eventually take steps intended to free their future populations from the dysgenic threat already so apparent. Responsible intellectuals were confident that morality would combine with science to save *Homo sapiens* from himself and enable him to take control of the evolutionary steering tiller and chart a rational course toward a happier future. Julian Huxley's prophecy seemed about to be fulfilled.

POSITIVE AND NEGATIVE EUGENICS

BUT CHANGES began to develop at this time within the eugenics movement which first seemed to strengthen the argument in favor of eugenic action, but eventually led to a major setback. Traces of this can be identified in the double-edged manner in which president Theodore Roosevelt had viewed the eugenics movement. Roosevelt strove mightily to persuade healthy American men and women to have more children, and fussed against members of the eugenics movement whose advice, he feared, could discourage childbearing among otherwise healthy Americans. Genetic science was not sufficiently advanced to permit the identification of deleterious genes, and Roosevelt and many other supporters of the eugenic ideal concluded that the best route to take in such circumstances was to encourage the procreation of individuals from families which had a record of fitness and creativity.

Theodore Roosevelt's criticism of eugenicists who were ignoring the international numbers competition proved to be a forerunner to a division which was to split the eugenics movement. Roosevelt might be described as being a "mainstream eugenicist." Mainstream eugenicists placed emphasis on positive eugenics, on encouraging the reproduction of healthy families chosen from among stock of proven historical capability. Being mostly of Northwest European origin themselves, the mainstream eugenicists saw eugenics as a national issue closely connected with the survival of their own stock at a time when the birth rate of that stock was declining and the population of non-whites was already growing with increasing rapidity.

But mainstream eugenics was doomed to lose the battle against what was called "reform" eugenics. As genetic science continued its initially slow and halting advance, many geneticists became so enthralled with the hunt for biological illnesses and defects that could be clearly identified as genetic in origin that they began to lose interest in mainstream eugenics – a goal which lent itself less readily to laboratory research. As genetic research advanced, geneticists became absorbed in the search for genetically-based diseases, and took little interest in encouraging creative families to increase or even maintain their birth rate – an activity which fell more realistically into the political arena. This search for inherited diseases was dubbed "reform" eugenics. Its goal could hardly be faulted; but as the memory of the massively dysgenic impact of World War I began to fade in the public mind, reform eugenics increasingly diverted attention away from positive eugenics, the ideal of those who sought to preserve a reserve of genetically healthy, intelligent and creative stock, in favor of reform or negative eugenics – the attempt to eliminate defective genes from being passed on to future generations. However, it would be many years before scientists could begin to foresee a day when it might be possible to eliminate individual defects.

Thus, the eugenics movement became divided at precisely the time when the eugenic ideal – the desire to engineer a healthy genetic heritage for future generations – came under increasing attack from those who were ideologically committed to egalitarianism. The latter refused to see the eugenic ideal in any light other than as an hierarchical concept implying superiority and inferiority – the precise pattern of thought they sought to eliminate from the social consciousness. They, too, sought to engage in social engineering, though engineering of a political nature which would have unanticipated dysgenic consequences, and the stage was set for the intense emotional struggle which today dominates both academia and the media, about the political correctness of permitting research into behavioral genetics, as well as the right to propagate information about the role of heredity in shaping the limits of human abilities and behavior.

Social Engineering and the Social Sciences

As we have seen, anthropology, sociology and the other social sciences were not formerly committed to egalitarianism and virtually all of these original pioneers recognized the role of heredity in determining human behavioral proclivities - many of them favoring eugenic research. Even Lester Ward condemned socialism as seeking to create an artificial equality, recommending instead the concept of a "sociocracy" which would benefit each according to his or her merit, but would provide equal *opportunity* to all. Ward supported the concept of eugenics, and recognizing the need for "race elevation" and "continuing the race",[8] he stressed the importance of women as progenitors of the race. Similarly, Thorstein Veblen warned against the inherent threat of "race suicide" and held that "providence is a virtue only so far as its aim is provision for posterity."[9]

The indisputable fact is that most pioneer social scientists were by no means egalitarians, and were quite open-minded so far as the possibility of incorporating biological knowledge into their altruistic plans to improve the human lot was concerned. The attempt by egalitarians ensconced in academe to deny the relevance of heredity to the human condition is of relatively recent origin, and has its roots in the

---

[8] Lester F. Ward, Outlines of Sociology, 1899, New York: Macmillan, pp. 147-149.

[9] Thorstein Veblen, *The Instinct of Workmanship,* 1914 NY: Macmillan

activities of radical political ideologues in the academic world who sought to advance the egalitarian thesis from respected positions of influence and authority.

From its inception the study of human behavior was strongly linked to the natural human urge to use knowledge to improve the human condition. Indeed, sociology is customarily traced back to Comte Henri de Saint-Simon (1760–1825) – one of those extraordinary French aristocrats who acknowledged a degree of corruption in the *ancien régime* and genuinely desired to create a "brave new world" – though one that differed substantially from that envisioned by Voltaire and Rousseau. Saint-Simon was an idealist who enrolled in the French army at the age of seventeen, fought with the French expeditionary force sent to help fight the British in the American Revolution, and was present at the surrender of Cornwallis at Yorktown. He believed that a scientific, positivist approach to the study of society could eradicate many of the evils that beset mankind; but not being an egalitarian, he did not support the ensuing French Revolution which had a far different character from that of the American Revolution.

Imprisoned during Robespierre's reign of terror because of his aristocratic background, Saint-Simon lived to survive this bloody period in French history and to devote himself to the question of social and political reform. His ideal was a hierarchical society organized around an efficient system of industrial and agricultural productivity such as that which had made it possible for Britain to stand up against the power of Napoleon. His proposals aimed at "ratonalizing" society caught the eye of Napoleon, who attempted to impose some of them on the new United Europe he was attempting to create. Saint-Simon hired a young secretary, Auguste Comte (1798–1857), who admired his employer's ideas so much that he decided to make them his own. Consequently, it was Auguste Comte, writing after Saint-Simon's death, who is commonly regarded as the founder of sociology.

Like his mentor, Auguste Comte conceived of sociology

as a science aimed at developing a system of efficient social control which would nevertheless be in harmony with man's biologically determined nature. His ideal was also an hierarchical system, with captains of industry holding the controlling positions; it was in no way egalitarian and it rejected the idea of biological egalitarianism. However, egalitarianism steadily gained strength as a result of the various revolutionary passions that swept Europe in the nineteenth century, and while some of the socialists who preserved a sense of national loyalty identified with the eugenic concept, most moved in the direction of egalitarianism, accepting Saint-Simon and Comte's advocacy of social engineering while rejecting the hierarchical concept of society implicit in more traditional philosophy. Notable amongst these was Franz Boas, who migrated to the United States from Germany to become the head of the first American Ph.D.-granting university department of anthropology. Boasian anthropology eventually became the bulwark of those in academe who were committed to the myth of biological egalitarianism.

V
RADICAL EGALITARIANISM PENETRATES ACADEME

JUST WHEN IT SEEMED that the nations of the Western world were rising to a new level of social consciousness and moral responsibility, World War I wrought political and genetic havoc upon the Western world. Striking Europe the most heavily, with Britain and France suffering genetic losses from which they never recovered, it also led to significant changes in America which were further accentuated by the ensuing Great Depression. The forces of revolutionary socialism, seducing significant segments of the public by painting a fanciful image of a distant egalitarian paradise, were the only beneficiaries of the disastrous maelstrom of destruction wrought by the war. Although eugenicists saw the war losses as additional evidence of the urgent need to preserve the drastically weakened heritage of beneficial genes, the public's confidence in the future was shattered by the Great Depression and the resultant struggle simply to survive. Popular concern, often little directed toward the distant future, was redirected even more strongly to the immediate problems, and the revolutionary socialist creed of egalitarianism steadily won control of the high ground.

In Russia, the massive losses of World War I led to violent revolutionary victory for those who preached anti-hereditarian views: although this brought with it only a cruel form of dictatorial government that showed no interest in either environmental conservation or human genetic conservation. Even in Western Europe and North America, where there was no violent Marxist seizure of power, the forces of social revolution made egalitarianism their credo.

This steady slide from idealistic concern for the future to crude appeals to the material interests of the present generation was facilitated by the widespread penetration of the academic world by advocates of social revolution, and the influence of the latter consequently grew rapidly during an era troubled by economic depression.

Franz Boas

FRANZ BOAS (1858–1942) was born in Minden, Germany, and grew up in a politically radicalized environment. Both his parents were radical socialists who were active in the 1870-71 revolutionary movement that swept Europe. Abraham Jacobi, his uncle by marriage, had actually been imprisoned after being found guilty of armed violence in Cologne during the revolution of 1848, after which he had emigrated to America, where he soon became influential in the early rise of radical socialism. Boas's family background was thus intimately interconnected with the revolutionary socialist movement, which also provided him with ready contacts in the United States.

Although lacking formal training in anthropology, Boas had studied some cultural geography in addition to his main doctoral subject, which was physics. What gave him an entrée to anthropology was a report he wrote on the Eskimos of Baffin Land in 1883 for the *Der Berliner Tageblatt,* a liberal newspaper. After serving as a docent in geography at the University of Berlin, he migrated to the United States in 1886 to take up a position at Clark University. It was here that the first U.S. doctoral degree specifically designated as being in the area of anthropology was awarded under Boas's supervision. Boas subsequently became the head of a department of anthropology established at Columbia University, where he trained and awarded doctoral degrees to numerous selected students. Equipped with the earliest American doctorates specifically designated as being in the field of anthropology, his students by default became the leaders and prime builders of academic anthropology in the

United States, rapidly establishing themselves as the arbiters of anthropological research, publishing and teaching in American universities.

Interestingly, as late as 1911, in his book *The Mind of Primitive Man,* Boas had admitted that:

> [d]ifferences of structure must be accompanied by differences of function, physiological as well as psychological; and, as we found clear evidence of differences of structure between races, so we must anticipate that differences in mental characteristics will be found.

However, Boas was shortly to reverse this position when he realized that the recognition of genetic forces conflicted with the goals of his egalitarian and internationalist ideology, which sought to demolish the unity and coherence of national units. Instead he began a massive campaign to undermine national and ethnic consciousness and "combat racism" in whatever form it might find expression. In particular, his *Anthropology and Modern Life* (1928) and *Race, Language and Culture* (1940) were devoted to downplaying the concept of heredity and undermining the eugenic ideal. The egalitarian ideology steadily gained ground as Boas's disciples and *their* pupils gained increasing influence throughout American academe. The spread of Boasian doctrines was further facilitated by the position of world dominance then enjoyed by the Western nations. Spurred by an ethical desire to shoulder "the white man's burden" in a shrinking world, many academics came to believe that mankind should now abandon the Darwinian struggle and treat the diverse subspecies of mankind as members of a single, international gene pool. This entailed a reversal of the evolutionary process which has for all time been firmly rooted in procreative competition, and was an ethical concept not shared by the non-Western nations, who adhered to more functional, self-promoting, competitive patterns of behavior.[1]

Thus, despite mounting evidence that genetic factors were of profound importance in determining human physical and mental behavior, the desire that biological egalitarianism *should* be true gained strength as human altruism was redirected away from the immediate group toward even the most disparate members of diverse breeding populations. The evolutionary trend toward greater speciation, already threatened by increasing population and migratory pressures, was now further vitiated by cultural factors rooted in an ideology which favored overall sapiens homogenization. The new radicals in U.S. social science found it convenient to downplay heritability; and Boas's earlier acknowledgment of human biological disparities was edited out of his 1938 edition of *The Mind of Primitive Man.*

Those to whom Boas awarded doctoral degrees in anthropology generally shared his ideologies and became prime disciples of egalitarian universalism. As the only holders of designated doctorates in anthropology, they were natural candidates to head up newly-formed departments of anthropology in other universities – where they, in turn, trained and awarded doctorates to a new generation of Boasian anthropologists. Much of the egalitarian bias in anthropology can be traced to the influence of this one immigrant of solid egalitarian-revolutionary credentials.

In his biographical work *Franz Boas* (1953, p. 65), Melville Herskovits, himself a student of Boas, publicly acknowledged that Boas's "political sympathies leaned towards a variety of socialism common among Nineteenth Century liberals." A cynic might add that the nineteenth century liberals with whom members of Boas's family were connected included some of the bomb-throwing variety! The records of the United States House of Representatives (1944, p. 9)

---

[1] For an opposing view on this issue see *A New Theory of Human Evolution* by Sir Arthur Keith, 1940. London: Watts & Co., and other writings by the same author.

actually attribute Boas with no less than forty-six communist-front connections.

Herskovits (1953) went on to confirm that:

> [t]he Four Decades of the tenure of [Boas's] professorship at Columbia gave a continuity to his teaching that permitted him to develop students who eventually made up the greater part of the significant professional core of American anthropologists, and who came to man and direct most of the major departments of anthropology in the United States. In their turn, they trained the students who . . . have continued the tradition in which their teachers were trained. . . .

Despite such an open admission by his pupil, friend and biographer, few today fully appreciate the extent of Boas's influence on the subsequent history of anthropology in America. Yet it is readily apparent when we look at the careers of just a few of Boas's pupils, all of whom sought to revolutionize Western thinking about social relations and to minimize the role of heredity in the shaping of human abilities:

MARGARET MEAD, often called "the mother of American anthropology," became the author of many influential but biased anthropological texts. Mead earned her M.A. and Ph.D. from Columbia in 1929 under Boas, and obtained an influential position as associate curator of the American Museum of Natural History. Her *Coming of Age in Samoa* (1928) was a particularly blatant attempt to undermine Western family traditions in favor of free sex, and has since been shown to have flagrantly misrepresented Polynesian customs.[2] She co-authored with Gene Weltfish a pamphlet

---

[2] Mead's prime claim to anthropological fame, the book she wrote as a graduate student about Samoan sexual practices, was eventually exploded by D. Freeman's 1983 study entitled *Margaret Mead and Samoa: The Making and Unmaking of an Anthropological Myth* (Cambridge: Harvard University Press). The

entitled *The Races of Mankind* (1940), which somehow came to be adopted by the War Department for distribution among American World War II military personnel, until it was withdrawn when it was judged to be communist propaganda. Its general thesis reflected the very core of Marxist-Leninist doctrine, extending the concept of "class exploitation" to include "race exploitation" and portraying race as nothing more than a cultural invention designed to keep the colored races in submission – or a conceptual artifact designed to promote a specialized form of class exploitation. Mead was officially cited as having numerous communist-front affiliations.

MELVILLE HERSKOVITS obtained his Ph.D. under Boas in 1927 and was subsequently hired by Boas as a research associate in anthropology at Columbia. Herskovits later became chairman of anthropology at Northwestern University, and a major power in the increasingly incestuous world of American cultural anthropology, in which virtually all the key positions were controlled by students of Boas. As the first granter of doctorates in an area specifically designated as anthropology, Boas had brought off a political coup by which his views not only dominated subsequent generations of American anthropologists but also inspired leftist thinkers in related branches of the social sciences throughout the English-speaking world.

ASHLEY MONTAGU, born in England as Israel Ehrenburg, changed his name several times, adopting not merely a set of

---

ensuing pyrotechnics were effectively documented in Hiram Caton's 1990 publication, *The Samoa Reader: Anthropologists take Stock* (Lanham, MD: The University Press of America). Although the egalitarian forces were unable to defend Mead from Freeman's charges after the fraudulence of her research was exposed, it is significant that some of the most powerful forces in contemporary American anthropology still strove to rescue what they could of her academic image in view of her historic importance as one of the main disciples of anti-hereditarian egalitarianism and Boasian anthropology.

Anglophone names – which would not have been unusual – but one of uniquely aristocratic flavor, Montague Francis Ashley-Montagu. For a surname he appropriated the name Montagu from one of Britain's oldest titled families, and for good measure reinforced this by hyphenating it to the name of Ashley, since hyphenated surnames conveyed an additional aura of social status in Britain.[3] Resenting the hereditarian views of Britain's leading anatomist, Grafton Elliot Smith, under whom he was studying, he dropped out of the University of London without getting his bachelor's degree, and disappointed by the failure of the 1928 General Strike in Britain, the new "Montagu" migrated to America where he was awarded a Ph.D. by Boas in 1936 – and later appointed chairman of anthropology at Rutgers University. Montagu's entire career was built around a bitter crusade against the work of respected scholars such as Carleton Coon,[4] who recognized race as a vital product of human evolution. An influential torrent of publications, such as his widely promoted book which he chose to call *Man's Most Dangerous Myth: the fallacy of race* (1942), all sought to dismiss the biological reality of race.

---

[3] Pat Shipman, in *The Evolution of Racism* (Simon and Shuster, 1994, pages 159-160), documents the history of his series of name changes, from Israel Ehrenberg to Moses Israel Ehrenberg, then to the aristocratic Montague Francis Ashley-Montagu, and finally to today's somewhat more modest Ashley Montagu. She also documents the opposition to the UNESCO statement by leading scientists of that day, and notes the "overrepresentation" of Montagu's associates on the panel that devised the 1950 UNESCO Statement on Race.

[4] Prominent during his life as America's leading physical anthropologist, Harvard scholar Carleton Coon has since been persistently attacked by egalitarians despite the fact that subsequent research in blood groups, DNA and other genetic markers has shown that his system of racial classification, based largely on skeletal data and other outward phenotypical characteristics, was remarkably accurate in revealing the degrees of relationship among the diverse peoples of the world. See the bibliography attached to this article for mention of some of his leading publications.

Montagu was the rapporteur primarily responsible for drafting the academically absurd, but politically sententious, United Nations Educational, Scientific and Cultural Organization's (UNESCO) 1950 Statement on Race, which sought to deny the significance of racial disparities.[5] This view was widely protested by scientists from many countries, notably by the pre-eminent French paleontologist Henri Vallois, director of Musée de l'Homme.[6] Not all the members of UNESCO's own social science department accepted the original United Nations Statement. Professor Georges A. Heuse, director of the UNESCO ethno-psychology department, condemned it, commenting that it was "conceived chiefly by Ashley Montagu – an anthropologist who conceives race a social myth – and by a group of sociologists substantially ignorant of anthropology" (1955, p. 379). Indeed, the Statement prepared by Montagu was so absurd that it was rewritten and republished in 1952, only two years after it was initially adopted. The revised Statement, although still largely reflecting universalist-egalitarian views (as was only to be expected in view of the international and political nature of the United Nations) conceded the role of genetic factors in human life and admitted that human groups which were distinguished by "well-developed and primarily heritable physical differences from other groups" might

---

[5] Another notable contributor to the fallacious UNESCO Statement on Race was the politically-biased Mexican anthropologist Juan Comas, who in 1961 authored a chapter entitled "Racial myths" which appeared in a book entitled *Race and Science* published by the Columbia University Press.

[6] See Henri Vallois, 1951. "UNESCO on Race." *Man*, no. 28. Vallois became one of the earliest members of the editorial advisory board of *The Mankind Quarterly*, publishing in that journal and continuing to serve on its board until his death. In his article in *Man*, Vallois asserted his unequivocal opinion that "the existence of races of Man is an uncontestable biological fact."

justifiably be called "races".[7]

It would not be irrelevant to note that in 1942 Montagu chose to lecture at the so-called School for Democracy, which was classified as a communist organization by the New York legislature, and that he is on record as stating that "Soviet Russia is the outstanding example of perfect management of ethnic group relations" – a claim that might be challenged by many ethnic groups who had the experience of living as minorities under communist rule.

RUTH BENEDICT obtained her doctorate at Columbia University and became a lecturer at Columbia under the patronage of Boas, having been retained by him to teach there, where she eventually became a full professor. Her book *Patterns of Culture* (1934), containing an introduction by Boas and a preface by Margaret Mead, sought to emphasize the role of culture to the exclusion of biology in determining human achievement. It became required reading in so many college courses taught by acolytes of the Boas tradition that it eventually sold over a million copies. Then in 1940 a high point in the Boasian war against the concept of race and heredity was reached with the publication of her *Race: Science and Politics,* which essentially argued that the study of race was "a vicious and brutal impediment to human progress, brotherhood, and understanding."

ISADOR CHEIN obtained his Ph.D. in 1939 under Boas at Columbia and became one of the prime court witnesses to give "expert" testimony in favor of school desegregation. Chein gave the Court only his *opinion* about segregation, as though that opinion, coming from him, was *ipso facto* "scientific," without attempting to provide any scientific evidence to

---

[7] Active consideration is currently being given to further amendments necessitated by genetic and biological realism, although it may be confidently expected that the political considerations which dominate the United Nations will prevent the adoption of a truly balanced revision.

support it.[8]

KENNETH B. CLARK, a Panamanian African-American, obtained his Ph.D. under Boas at Columbia in 1940 and similarly became an "instant expert" on the American school system. Clark's testimony before the Supreme Court played an important role in determining the decision to force the racial integration of American schools in the landmark 1954 *Brown vs. Board of Education* decision (347 U.S. 483). Central to his testimony was his claim that because seven out of sixteen black children in an all-black school, when shown white and black dolls or drawings of such dolls, chose the white dolls as those "which most looked like them," segregation had "harmed the development of their personalities." The speciousness of his testimony was revealed by Ernest van den Haag, of New York University and the New School for Social Research, who showed from Clark's own data, collected in a study involving 253 black children from both segregated and nonsegregated schools, that a higher percentage (39%) of black children from racially *integrated* schools chose the white doll than did black children from segregated schools (29%).[9] In other words, Clark had mislead the Court with data that appeared to support his contention, while the data from his larger study showed the reverse of what he presented to the court.

OTTO KLINEBERG obtained his Ph.D. in psychology at Columbia University, where he also studied anthropology under Boas. He is credited with having helped persuade Boas

---

[8] See the account by Cahn, in "Jurisprudence", 31 *N.Y.U.L. Review*, 1956.

[9] In addition, more black children in northern (integrated) schools rejected the black doll than did black children in southern schools. See Ernest van den Haag, "Social Science Testimony in the Desegregation cases – a Reply to Professor Kenneth Clark," *Villanova Law Review*, Fall, 1960, 6:1, pp 69-79. Data is based on Clark's "Racial Identification and Preference in Negro Children," *Readings in Social Psychology*, pp. 174-5. Newcomb and Harley, Eds., 1947)

to suppress his former belief that race and intelligence were linked. After receiving his Ph.D. in psychology, Klineberg was promptly hired by Boas as a Columbia research associate in anthropology, and became noted for his persistent attacks on "hereditarian" views. In particular he wrote a chapter on "Race and Psychology" which was included in the politicized UNESCO publication entitled *The Race Question in Modern Science* (1956), and sought to explain recorded racial differences in intelligence as the product of purely environmental forces. His arguments were demolished by Henry E. Garrett,[10] who chaired the Psychology department at Columbia between 1942 and 1956, and served as President of the American Psychological Association as well as President of the Eastern Psychological Association and the Psychometric Society.

LESLIE SPIER was notable primarily for his constant support of the egalitarian claim that there was no evidence of any race being inferior in its ability to participate in any culture or even to create any culture – a sweeping statement which denies any validity to the concept of the heritability of behavior. The differing cultural achievements of races and populations were supposedly due solely to environmental circumstances and never to inherited differences in personality, intelligence, or any other forms of ability.

GENE WELTFISH obtained his Ph.D. from Columbia under Boas in 1929 and, after further studies under Alexander Goldenweiser at the New School for Social Research, where Margaret Mead also underwent postgraduate training, was hired by Boas to teach at Columbia. It was he who co-authored with Margaret Mead the notorious pamphlet *The Races of Mankind* (1948), mentioned above.

---

[10] See "Klineberg's Chapter on Race and Psychology" by Henry E. Garrett, in Volume I. No. 1 of *The Mankind Quarterly*, July 1960.

This is a remarkable and formidable list of politicized scholars, all educated under the same man and all reflecting his moral commitments. The result has been catastrophic. As Austin L. Hughes wrote in the final chapter, entitled "Prospects for a Biologically Based Social Science", of his 1988 book *Evolution and Human Kinship:*[11]

> ... the deficiencies of American anthropology as a serious science of society may be at least partly to blame for the seeming inability of United States policymakers to understand or predict accurately the effects their policies will have .... If policy-makers in the United States get their ideas of how societies function from undergraduate readings of Mead, Benedikt, and Schneider, it is small wonder that they have difficulty in dealing with the rest of the world.

The collective influence of these politicized academics within American academe early became obvious to those who refused to close their eyes to the biological basis of human behavior, and it has been said that the Galton Society regarded the need to combat Boasian egalitarian and extreme environmentalist propaganda as one of its chief goals. But since the Boasians dominated academe as official faculty members, working together to seize control of key positions and recruiting into academe only those of their students who had accepted their views, the members of the Galton Society were outgunned and outmaneuvered, and fought a losing battle. At this point those who still recognized the importance of genetics found themselves confronted by a new ideology, neo-Marxist, Stalinist Lysenkoism.

---

[11] New York: Oxford University Press, p. 139

VI

EGALITARIANISM ASCENDANT: MARXISM AND THE THIRD WORLD POPULATION EXPLOSION STRENGTHEN THE ANTI-HEREDITARIAN CULT

SCIENTIFICALLY, the spread of egalitarian theory was encouraged by the emphasis placed on the environment by the behaviorism of J. B. Watson (1878–1958). This expanded Pavlovian research into a broad philosophy which assumed that human beings were essentially plastic organisms equally able to develop in virtually whatever direction might be determined by their environment and schooling. Watson, who spent a good part of his adult life as an executive with large advertising firms such as J. Walter Thompson, encouraged researchers to account for sociopsychological phenomena in environmentalist terms. This led many to overlook the existence of significant differences in the heredity of individuals. In short, an extreme environmentalism developed which assumed that all healthy human beings were equally capable of responding to training in any direction. This view was extended into the field of education by John Dewey (1859–1952), a pragmatist who believed that environmental manipulation could reshape human behavior in virtually any desired direction and whose influence on American education persists to this day. Such simplistic behaviorism rejected the idea that biology affected either individual or group human behavior.

The importance of environment had never been ignored in European thought, and indeed traditional educators had always placed great importance on a strict regime of training and schooling; the older tradition differed from the new behaviorism in that it recognized that heredity placed limits on the individual's ability to benefit from experience and from educational opportunity. By contrast, the new emphasis on environment, to the exclusion of all biological factors, proved a fertile field for those whose political agenda depended on promoting egalitarianism.

Consequently, the egalitarians perceived that their most fertile field of influence lay in the quasi-scientific realms of sociology and anthropology, where their theories could be advanced without being subject to experimental test. These areas increasingly attracted egalitarian attention, and soon new generations of scholars were being trained and graduated in that tradition. Physical anthropology was cut off as something totally separate from cultural and social anthropology, thus allowing the latter field to develop a purely environmentalist approach which left students ignorant of the possibility of human biological differences.[1]

The economic depression that followed World War I created a political atmosphere which facilitated the promotion of the myth of biological egalitarianism, and the events of World War II weakened opposition to that myth. Biological egalitarianism now became the "scientific" underpinning for the environmentalist conceptions of society which spread throughout both the social sciences and the media. Political pressures swung major foundations which funded both educational research and socio-political programs into alignment with the egalitarian creed, and a vast body of pseudo-science rapidly came to dominate the fields of anthropology, sociology and philosophy, with biased academic opinions about biological egalitarianism even influencing

---

[1] One of the major achievements of the Boasian environmentalist cult was the detachment of "cultural" anthropology from "physical" anthropology as though the two were quite different and unrelated sciences and that environment and culture superseded all genetic and biological qualities. For an approach to anthropology as an integrated science of human behavior involving biological as well as cultural evolution, see Part I "The Evolutionary Background" in Roger Pearson, *Introduction to Anthropology* (1974), New York: Holt, Rinehart and Wilson.

Court decisions and governmental policy relative to social issues.

MARXISM ADOPTS THE LYSENKOVIAN MYTH

THE BOASIANS were not the sole source of anti-hereditarian views in academe. The idea of the *biological* equality of all mankind was a virtual necessity to political egalitarianism. In Russia, a quack geneticist named Trofim Lysenko rejected Mendelian genetics and claimed that the environment could directly modify the genome. This fallacy, linked to the equally erroneous Lamarckian doctrine that adaptations acquired during the lifetime of individual organisms could be inherited by that organism's offspring, reduced breeding to philosophical irrelevance, thereby promoting the cause of egalitarians and those who believed they could reshape mankind at will by means of environmental manipulation without the need to resort to any form of genetic selection. The Lysenkovian theory was adopted by Stalin, who declared Mendelian genetics to be a capitalist conspiracy, and acceptance of the Lysenko fallacy became mandatory throughout the Soviet Union. Lysenkoism justified the doctrine of revolution by consigning Mendelian genetics to oblivion: if all races and individuals were potentially equal, given the appropriate manipulation of their environment, the argument that all economic inequality was the product of class exploitation would be logically justifiable.

With the advent of the Lysenko doctrine, revolutionary Marxists could forget Marx and Engels' belief in racial differences and join the anti-hereditarian chorus with enthusiasm. Even outside the Soviet Union, Marxists now joined non-Marxist egalitarians in attacking all and any scholars who adhered to Mendelian theory, especially when these persisted in drawing attention to the link between genetics and human ability. In Britain, the ardent eugenicist J. B. S. Haldane resigned his Communist Party membership in protest, and in America the University of Texas Nobel Prize-winning eugenicist, Hermann J. Muller, who had long been a card-carrying member of the CPUSA, abandoned his membership after narrowly escaping arrest during a visit to the USSR. His crime: he had publicly condemned the Lysenkovian myth as unscientific.

Lysenkoism remained for many years the only legally permitted form of genetics in the USSR, and most Marxists ensconced in the social science divisions of Western universities accepted the Lysenkoist doctrine out of political conviction. The story is well-documented in Conway Zirkle's illuminating book, *Evolution, Marxian Biology and the Social Scene* (1959).[2] Biological egalitarianism became the only "politically correct" doctrine among Marxists throughout the world, and Zirkle shows how widely what he calls "Marxian biology" permeated Western anthropology, sociology, and related studies through the teachings of faculty members who were ideologically attracted to egalitarianism but were balefully ignorant of even elementary biology. To quote his words:

> Marxian biology ... exists also in non-Communist countries – in countries where it is not protected by Marxian dictators. Moreover, it exists not merely as an intellectual lag among the unlearned, but as a carefully protected faith in disciplines whose members are equal in education – quantitatively at least – to the biologists themselves. (1959, p. 418) ... The usual course is to treat the human species as if it were composed of an amorphous, uniform and plastic raw material, as if it were a species which could be molded (conditioned is the usual word) to suit the heart's desire." (p. 420)

All kinds of eugenics are anathema to Marxists, Zirkle affirmed, even though already in 1959 our knowledge of the machinery of hereditary was sufficient to enable him to assert that any rational eugenic program would be beneficial to

---

[2] See also Jamieson (1982).

mankind. Indeed, the intellectual climate was so bad that the Marxists ensconced in Western academeme

> ... actually set the fashion not only in letters but also in popular up-to-date attitudes in morals and ethics .... It is even possible that they furnished the dominant directives to the social sciences. This statement is not as far-fetched as it might seem at first, for practically all social scientists are familiar with the works of the more progressive writers, but almost none of them is technically equipped to evaluate the new discoveries in biology.... Marxian biology has always had allies, and this has been one of the sources of its strength. On the other hand, scientific biology has had few friends. The moment it grew to the point where it applied to *Homo sapiens*, it acquired enemies."(p. 298)

Shattered by the deeply dysgenic impact of World War I, and by the economic depression that followed, many who had previously put loyalty to the national ideal ahead of selfish interests now fell prey to the egalitarian myth, correctly blaming the members of the old elite for leading their countries into that disastrous conflict. Egalitarianism portrayed any ranking of human talents as nothing more than an artificial justification for class exploitation. The idea of eugenic policies, based on a recognition of the reality of biological inequality, of competitive reproduction, and on the fear that modern social conditions negated natural selection, now became increasingly disturbing to those who desired only to live their own lives free from worries about the wellbeing of future generations.

The earlier list we cited of radical social scientists was restricted to some of the more prominent of Boas's pupils.[3]

---

[3] But the heirs of the Boasian school remain ensconced in the social sciences, and some ideologues have recently infiltrated even such a respected organization as the Behavioral Genetics Association, demanding the resignation of the Association's then president, professor

However, the Boasians were now joined by other blatantly radical academics. Gerhard Lenski and Marvin Harris, for example, advanced an essentially Marxist analysis of social behavior rooted in the concept of egalitarianism and class warfare which exerted a major influence in American sociology.[4] In addition, a bitter campaign of guerilla warfare against those who acknowledge the role of heredity has been carried out for years by associates of an activist radical publication known as *Science for the People* – which uses the term "people" in its most traditional Marxist interpretation, reminiscent of the "Hall of the People" in Beijing.

The political left-wing had now achieved ascendancy in the universities of the Western world. Indeed, many contemporary evolutionary scientists still seem to wish to be perceived as believing in equality rather than in hierarchy, and in a degree of malleability of human nature that does not exist. Evolutionary scientists consequently focused their writings on the "panhuman" traits shared by all living hominids, and emphasized the concept of "facultative adjustment" to show how people altered their behavioral strategies dependent on circumstance. They went out of their way to minimize any view that the people of Japan, for example, were different from the people of Jamaica, except as a result, perhaps, of their differing economic or social arrangements.

The outbreak of World War II and the ensuing defeat of Germany made it possible for the egalitarians to smear all who sought to advocate eugenics as "Nazis", "anti-Semites" and "racists". World War II gave the political edge to the

---

Glayde Whitney, after he gave a talk suggesting the need to investigate the possibility of genetic factors behind the high incidence of black crime in America (see G. Whitney, "Ideology and Censorship in Behavior Genetics" in *The Mankind Quarterly*, Vol. XXXV, No 4, Summer 1995).

[4] Lenski's *Power and Privilege* (1966), a widely used textbook, presented an outright Marxist view of society in that it sought to examine human cultures entirely in terms of class exploitation and stratification.

anti-eugenicists. Germany had adopted the same eugenic policies as were in vogue throughout most of the advanced countries of the Western world, and information about this became intermingled with the holocaust reports coming out of post-war Germany. The outcome was that all discussion of eugenics was illogically but effectively equated with inhumanity – a deliberate inversion of the highly moral position that prompted most eugenicists to support the movement.

The egalitarians, of course, sought to argue that the idea of eugenics promoted the more murderous instincts in humankind, never mentioning the historical fact that egalitarian enthusiasm had also historically resulted in totalitarianism – and in the greatest genocide the world has ever seen: witness the mass slaughter of Russian "bourgeois" peasants, the Katyn Woods massacre of Polish officers, the uncounted deaths in Siberian "gulags", and the even more massive genocide of the cultured and landowning classes in Maoist China – all in the name of "equality." Ideological bias prevented the egalitarians from seeing or even desiring to see objective reality.

Irrational political ideologies can severely retard the progress of knowledge, but in a free country they cannot suppress the search for truth altogether. While the Lysenko tradition still raises barriers to research into behavioral genetics, genetic research is now once again advancing rapidly, and increasing numbers of psychologists and social scientists are examining the role of genetics in human behavior, even though their interest still draws passionate condemnation from those who retain a Stalinist-like commitment to the myth of biological egalitarianism. Seeing their cherished notions challenged by research into human heredity, these latter still strive with a vigor born of desperation to suppress public recognition of the importance of heredity.

NEEDED: "A TRUE SOCIOLOGY OF KNOWLEDGE"

TO UNDERSTAND the history of academic fashion one

cannot ignore the influence of political interests, since these exert a powerful influence on academic ideologies at any one time. Thus, Karl Mannheim, a Marxist immigrant who fled to Britain from Germany, was certainly right in advocating the need for a "sociology of knowledge" (1936) on the grounds that scholarly research is influenced by the personal conceptions, values and politics of the researcher. Unfortunately he saw this issue only in a left-wing light, attributing all non-Marxist theories to bourgeois class bias, and so hindered rather than advanced the development of this branch of knowledge.

An analysis of the biases and political convictions of those who teach in the classrooms of our schools and universities would go far toward explaining the academic theories which they expound. Reason tends to serve the emotions, and it is the emotions that drive human curiosity and human activity. This does not mean that some scholars are not truly emotionally driven to understand the real world. Our early forebears must always have felt insecure when faced with a situation they could not understand, and today a true scholar feels uncomfortable when faced with a problem he cannot solve. But unfortunately facile theories such as Marxism can appear to explain all facets of human behavior. Marxism takes its cadets to the top of a mountain and shows them how the world looks from that single viewpoint, and those with less inquisitive minds are too often satisfied with the map that then appears to unroll before them. But there are many intellectual viewpoints or mountain tops from which reality can be observed, and to believe that every facet of human behavior can be explained in purely economic terms is overly facile and academically ridiculous.

The Third World Population Explosion

But there is another force at work promoting a public willingness to accept the myth of biological egalitarianism. In *Darwinism Applied*, John H. Beckstrom shows how evolution among the more advanced species resulted from competition between disparate populations, with ethnic loyalty tending to cause individuals in a natural environment to promote the survival chances of those who are closely related to themselves and therefore endowed with a similar collection of genes. From a genetic viewpoint, he observes, all that is important is that the genes survive, for "all the identical copies of a gene in whatever bodies they appear are of equal value to the genetic material they represent." (1993, p. 15). It is the gene pool of the kinship unit that is the key to understanding evolution.

The evolutionary function of altruism is frustrated and the sense of group identity is weakened in multi-racial societies. Thirty-three years ago, Sir Julian Huxley pointed to the perversion of altruism that has resulted from the rise of large urbanized, multi-racial communities, and the dysgenic result of this re-direction of a natural impulse into unnatural channels in an overcrowded and heavily acculturated world:

> In that long period of human history during which our evolving and expanding hominid ancestors lived in small and tightly knit groups competing for territorial and technological success, the social organization promoted selection for intelligent exploration of possibilities, devotion and cooperative altruism: the cultural and genetic systems reinforced each other. It was only much later, with the growth of bigger social units ... that the two become antagonistic ... and gave way to the possibility and later the probability of genetic regression and degeneration.[5]

Biological egalitarianism is being further advanced by the psychological impact of the Third World population explosion. World population is currently increasing at the rate of some 97 million annually, and even assuming that the rate of increase will decline, it is estimated to reach 10 billion by the

---

[5] Sir Julian Huxley, "Eugenics in Evolutionary Perspective". *The Eugenics Review,* October 1962, p. 133.

year 2050. This is prompting massive migratory movements, which are probably only a token sample of what is yet to come. The population of China, where strict attempts are being made to limit population growth, is currently 1.25 billion, but even assuming that government efforts to restrict population growth continue, the Worldwatch Institute projects a Chinese population of 1.66 billion by the year 2045. Women in Africa have for some decades been averaging six live births per female, whereas in areas like northern Italy and east Germany the birthrate is far below replacement level. The population of Bangladesh (formerly East Pakistan) was around 40 million when the area was first separated from India only 48 years ago; today it exceeds 125 million. Needless to say, Bangladesh supports this population increase only as a result of massive aid from the countries of the Western world, whose share of the world's population is annually declining. The population of India increases every five or six days by the total population of Europe's oldest democracy, Iceland. The changing character of the human population is also reflected in disparities in birthrates between the more successful industrialized countries and the less successful Third World peoples, who worldwide, on average, are doubling their populations every quarter of a century. The population of native Europeans is actually declining, and it is anticipated that Europeans will eventually be largely replaced by increasingly massive immigration from Africa, Asia and the Far East. The same is also likely to happen to the European-descended populations of Australia, New Zealand, and even North America.

Genetic Altruism

Egalitarian propaganda was consequently powerfully aided by a separate development that now began to pervade segments of academia, the media and the political world. In the nations of the West, already quasi-internationalist due to the legacy of Christianity as a missionary religion, modern communications tended to extend the concept of human community to embrace the entire world. As a result, sympathy for the economic plight of the burgeoning masses of the Third World – due to the introduction of modern Western medical technology – led to a substantial redirection of altruism away from the welfare of the immediate gene pool to engulf the most diverse members of the hominid species.[6]

This new spirit of biological egalitarianism involved a transmutation of humanitarian values. Altruism had evolved because the survival chances of a population were enhanced when individual members were prepared to make personal sacrifices to protect the intergenerational well-being of their own kind. Today, however, altruism is becoming increasingly generalized among the peoples of the Western world, who are losing their sense of ethnic cohesiveness and distinctive identity. For them, altruistic emotions are becoming increasingly stimulated not by threats to the survival of their own gene pool, but by the plight of all the world's expanding multitudes, which in past millennia nature had always trimmed back by natural selection.

As a result of this redirection of the altruistic impulse, the members of the advanced Western nations increasingly devote their efforts to attempts to assist genetically competing populations to survive and reproduce prolifically, while

---

[6] To understand the wide range of disparate qualities which distinguish men from each other today, it is necessary to reflect upon the evolutionary history of our species. Earlier hominids evolved as a variety of widely differing subspecies which, in the normal course of evolution, isolated in widely dispersed regions, would eventually have evolved into quite separate species – but for the fact that the more advanced populations successively tended to expand into neighboring territories, either supplanting or admixing with the earlier occupants of those territories. As a result, today's population of *Homo sapiens* comprises the descendants of a more advanced subspecies whose forebears mixed in remote areas with the remnants of older local subspecies. Through such local genetic admixtures, the diverse populations and peoples of the modern world were created. By supplanting the more primitive hominid subspecies, and absorbing others, *Homo sapiens* remained one species, capable of fertile crossbreeding among its various subspecies and mixtures of subspecies, characterized by the wide disparity of innate biological properties which distinguish both living individuals and races from each other.

allowing their own birthrates to fall below replacement level. As Garrett Hardin has pointed out repeatedly, and most notably in *The Limits of Altruism: An Ecologist's View of Survival* (1977), this redirection of the altruistic impulse is effectively standing evolution on its head.[7]

The term "genetic altruism" has been used to describe the actions of those who willingly refrain from passing on their own genes to posterity while assisting other individuals and groups to reproduce, many of them prolifically. This inversion of the evolutionary function of altruism was accompanied by a decline of interest in eugenics. A new morality was developing which demanded that advanced societies should fight against evolution: natural selection was to be replaced by dysgenic selection – nature was to be prevented from wielding its genetic pruning knife, and no attention was to be paid to the well-being of all those future generations of mankind who would be doomed to eke out their existence in an increasingly overcrowded world blighted by man-made dysgenic population trends.

Although this new anti-evolutionary philosophy of life took hold primarily in the more advanced countries, where the birthrate began to fall dramatically, it has so far been widely rejected among the peoples of the Third World. By adhering to more traditional reproductive practices, and benefitting from Western economic and medical aid, these populations began to explode at geometric rates leading to demands for yet more aid and larger transfers of wealth from the donor countries.

Evolutionary progress is rooted in the process of speciation and competition between populations and

[7] Garret Hardin, 1977, Bloomington: The Indiana University Press. However, see also other works by Hardin listed in the bibliography. The work of University of Western Ontario psychologist, J. Philippe Rushton, who explains how finely tuned are people's discriminations based on genetic similarity, even within the family, is also relevant here. Rushton provides a theoretical model showing how group selection can occur. See "Genetic Similarity, Human Altruism, and Group Selection" in *Behavioral and Brain Sciences*, 1989, 12:503-559.

subspecies to reproduce and control vital resources. Altruism served an evolutionary function among advanced social animals – the enhancement of the survival chances of their own phylogenetic continuum. When the breeding population was itself not only the community but also the total society this worked well. But modern communications and modern mass migrations have reversed the evolutionary impact of altruism, commonly redirecting it away from members of the immediate gene pool to aid the members of competing gene pools to reproduce at the expense of the altruist's own genetic heirs. Humane though it may seem to us, the contemporary Western trend to redirect altruism away from their own genetic kinfolk toward competing populations of different genetic stock represents a reversal of the evolutionary process, and could potentially result in disastrous consequences. This is all the more true because the altruistic extension of modern medicine to all corners of the earth has created the threat of world overpopulation.

Today, Marxists and other biological egalitarians, somewhat humbled by events in Eastern Europe, but with their ethical arguments reinforced by the changing attitudes of the Western nations as a result of the new multiculturalism arising from the presence of massive immigrant minorities, remain dis-proportionately influential in the universities and media of the Western world. Large numbers of social science academics still refuse to recognize, or lack the courage to openly recognize, the implications of the fact that human beings are biological organisms, even though an increasing majority of psychologists once again acknowledge the importance of heredity in shaping the limits of human personality and abilities.[8] The result has been increasing intimidation of those who seek to apply Darwinian logic to the solution of human problems.

---

[8] See Snyderman and Rothman, *The IQ Controversy: The Media and Public Policy*, 1988. Transaction Books.

VII
MEDIA AND ACADEMIC PERSECUTION OF "HEREDITARIAN" SCHOLARS

IN THE EARLY decades of the present century the political right had persuasively argued that evolutionary theory supported their social philosophy, but the political left had then come to realize that the concept of biological inequality and "survival of the fittest" was incompatible with the notion of equality. Opposition to German National Socialism clearly played a significant role in blunting Darwin's impact. From the 1930s onward, scarcely anyone outside of Germany and its Axis allies dared to suggest that human populations might be in any respect genetically different from others lest it should appear that the author held "racist" views and supported the National Socialist cause.[1]

Meanwhile, those who propounded the myth of biological egalitarianism were free to write what they liked, without fear of contradiction. They made full use of their opportunity in the decades that followed. Politically fuelled also by European decolonization, by the onslaught against apartheid in South Africa, and by the U.S. civil rights movement, the idea of a genetically based core of human nature on which individuals and social groups might differ significantly was consistently derogated.

As Earnest Albert Hooton observed in 1941, just two years after World War II had erupted in Europe, there was

---

[1] Even today anyone who so much as suggests that there are racial differences in intelligence is likely to be called a "racist" or even "Nazi" by radical political activists, even when it is a notable scientist who fled from Nazi Germany (e.g. Hans Eysenck: see his autobiographical remarks in Pearson 1991), or Jewish scholars such as Herrnstein, 1994.

now a growing reluctance to recognize the reality of race. In his own words:

> ... in this country [the U.S.A.], which is a welter of races and transplanted nationalities, one can assert without much fear of contradiction and with every expectation of applause that all races are equal in ability and in cultural achievement and that there is no relation between race and culture. No one's ox is gored thereby and nobody bellows. But if you suggest to the contrary, every ethnic strain with an inferiority complex becomes vociferous in protest. (1941, p.103)

The ideology of egalitarianism at this time became so strongly entrenched in the universities of the Western world, that many contemporary evolutionary scientists still seem to exhibit a pervasive wish to be seen as supporting the idea of the biological equality of living hominids, rather than to admit the extent of biological differences which characterize living individuals and populations. A belief in an extreme and unreal malleability of human nature became the orthodox view, and even evolutionary scientists focussed their attentions on the "panhuman" traits shared by all living humankind, and emphasized the concept of "facultative adjustment" showing how hominids readily altered their behavioral strategies to suit environmental conditions. This was all very well, but it became mandatory to minimize any evidence that the people of Japan, for example, were biologically different from the people of Jamaica. All differences were attributed solely to differing economic or social arrangements.

During the 1960s and '70s, there was openly violent intimidation by Marxists and other leftist activists of scholars who refused to ignore the genetic component behind human behavior. Marxist SDS-organized[2] campus riots aimed at "racist" professors were frequent, as documented in *Race, Intelligence and Bias in Academe* (Pearson, 1991) – and in more specific detail concerning the harassment of the Nobel Prize-winning William Shockley in *Shockley on Eugenics and Race* (Pearson, 1992).[3] Leftist student organizations and their sympathizers, encouraged by like-minded faculty members, rioted on campuses, disrupted lectures, threatened faculty members, and attacked the right of "hereditarian" scholars to report their research findings or to express their opinions either on *or* off the campus.

During the 1980s, a degree of order returned to academe, but by now many of those who had been student activists in the 1960s and '70s had followed the course advocated by the Italian Marxist Antonio Gramsci (1977), and had abandoned violent revolutionary activities in favor of infiltrating the established institutions of society as paid employees. Numbers of the rebellious students of the 1960s and '70s became the faculty members, university administrators, and government bureaucrats of the late 1980s and '90s. Whereas universities such as the University of California, Berkeley, had hitherto defended the right of faculty members to express their opinions, numerous universities and other educational institutions now became leaders of the egalitarian cause, imposing restrictions on the freedom of faculty members to engage in research or practice free speech wherever these involved the influence of genetic factors on human behavior. Guidelines were laid down as to "racially sensitive" and

---

[2] The ludicrously named Students for a Democratic Society (SDS) was an outright Marxist organization whose activities are detailed in *Far Left of Center: The American Radical Left Today* by Harvey Klehr, Transaction Books, New Brunswick. 1988.

[3] Both of these books – Roger Pearson, *Race, Intelligence and Bias in Academe* (1991), with an introduction by Hans J. Eysenck, and Roger Pearson, *Shockley on Eugenics and Race* (1992), with an introduction by Arthur Jensen – are available from Scott-Townsend Publishers, Box 34070, NW Washington DC 20043-4070.

"politically correct" terminology at Pennsylvania State and other universities and faculty members who published papers on the genetic basis of personality were discriminated against at universities such as Western Ontario, Delaware, and City University of New York.[4]

As the doctrine of biological egalitarianism gained in political and emotional support, it began to inhibit not only individual scholars but also grant-making foundations. Even foundations which had been established by conservative and heredity and eugenics-oriented benefactors, such as the Ford Foundation, the Carnegie Institute, and the Kellogg Foundation, began to eschew support for research that might favor genetic rather than exclusively environmental analyses of human problems. For a broad account of the persecution of scholars who rejected pure environmentalism refer to Pearson (1991), mentioned above,[5] but this does not cover the numerous instances which have occurred since that date, notable amongst which is the cancellation of a $78,000 grant from the National Institutes for Health to support a University of Maryland conference on genetics and criminal behavior, under pressure from political groups, particularly the U. S. Congressional Black Caucus and the National Association for the Advancement of Colored Peoples (the latter being largely funded by massive tax-exempt organizations such as the Ford Foundation). The conference, originally scheduled for 1992, finally took place in 1995 under heavily restricted guidelines.[6]

Unable successfully to refute solid research into the significance of genetic forces in determining the limits of human behavior, the egalitarians frequently resort to (1) accusations that individual researchers utilized poor methodology or even outright fraud, such allegations relying

---

[4] For details see Pearson 1991.

[5] See footnote 49

[6] See Joyce Price, "Crime-and-genes conference resurfaces with new focus" *Washington Post*. September 22, 1995.

on the principle that when mud is thrown, some at least will stick, and (2) the argument that any research into the role of genetics in determining human behavior will lead inevitably to "discrimination" and therefore is morally unjustifiable. Currently, there is considerable publicity aimed at research into the genetic basis of breast cancer, on the grounds that this will lead to "discrimination" by insurance companies and employers. What the media should be discussing is the vital importance of medical genetic research, the benefits that mankind can obtain from knowledge about genetic causality, and the fact that as medicine advances there is an increasing possibility of developing a humane eugenic program, based on the development of genetic surgery, which could help free future generations from some of the ills which plague the present generation.

Many religious authorities, immersed as they are in the concept of divine predestination in respect of man and the universe, speak out against any form of eugenics. Few have the courage, or inclination, of Rabbi Lord Jakobovits, former chief rabbi of Great Britain, who recently endorsed eugenics when he went so far as to say that he would favor genetic engineering as a means to solve social problems such as homosexuality, if that were found to be scientifically feasible.[7] This observation was unfortunately socially tendentious, and is the kind of proposition that causes many to recoil from consideration of the possibilities that genetic knowledge now offers humankind. Yet the use of genetic engineering to eliminate serious genetic illnesses, and to ensure that an adequate supply of intellectual prowess is passed on to future generations of humanity, should surely not provoke the hostile reactions it so often raises amongst those who seemingly care but little for posterity.

Media Misrepresentation

---

[7] "Rabbi Favors Genetic Engineering," *The Detroit Jewish News*, July 30, 1993.

AS HAS BEEN revealed by Snyderman and Rothman in *The IQ Controversy: The Media and Public Policy* (1988), these radical ideologues generally found willing collaborators in the media. The intimidation of "hereditarian" scientists became a major plank in the politics of "political correctness."

Politicized distortion of scientific findings has not been limited to the popular media. Sadly, a number of scientific journals fell under the control of individuals who minimize the role of heredity in shaping human behavior. Many of these journals are influential because of the remarkable service they provide in spreading scientific information. The *Scientific American* is one such publication which, despite its general excellence, sympathetically provides a platform for such far-left writers as Leon Kamin, the former New England editor of a Communist party weekly newspaper, whose views were recently further promoted by staff writer John Horgan in a subtly biased article entitled "Eugenics Revisited" (June, 1993). This was prominently advertised on the cover of the issue in which it appeared, as exposing "The *dubious* [our italics] links between genes and behavior".[8] Amongst other evidence of bias, Horgan gave considerable space to Kamin's erroneous critique of the Minnesota Twin Study.[9] In the January 1995 issue, *Scientific American* staff writer Tim Beardsley contributed another distorted antihereditarian piece, this time attacking Herrnstein and Murray's massive

---

[8] Contemporary scientific research clearly establishes the authority of animal behavioral genetics, and no geneticist would ever consider alleging that human beings were not animals, and that the laws of causality do not apply to human beings.

[9] Journals and newspapers that publish distorted or erroneous material too often ignore letters of refutation. Berkeley anthropologist Vincent Sarich wrote a letter protesting Horgan's article, which was never published. Furthermore, when they do agree to publish letters of refutation they customarily allow the original author space to respond, while denying space to the complainant to refute new distortions contained in the "response."

study, *The Bell Curve* (see later) with the journalistic title, "For Whom the Bell Really Tolls". To this title the *Scientific American* editors chose to publish an additional subtitle: "A tendentious tome abuses science to promote far-right policies".

Kamin, the author of heavily politicized books such as *The Science and Politics of I. Q.* (1974), in which he states that "[f]or all we know the heritability (of IQ) may be zero ...", and *Not in Our Genes: biological ideology and human nature* (1984),[10] participates in personal attacks against those who attempt to draw the attention of the public to the role of heredity in human behavior. He adheres to the dream of human biological egalitarianism and has attacked "hereditarian" scholars in articles and letters which on occasion contain such flagrantly misleading statements that one can only suppose that he places an unscholarly reliance on secondary sources which he has not troubled to check for accuracy. It was Kamin who claimed credit for disclosing fraud in the publications of Sir Cyril Burt – the prominent British psychologist who early proclaimed the largely genetic basis of intelligence – on the grounds that Burt's data could not be validated. What he did not mention was that following Burt's death, Burt's twin studies research records were reportedly destroyed on the recommendation of an unsympatheticcolleague. Meanwhile, Burt's published findings on the intelligence of twins have since been replicated by other researchers and the charge that Burt manufactured his data has been examined and refuted by Joynson, in *The Burt Affair* (1989), and by Fletcher in *Science, Ideology and the Media: The Cyril Burt Affair* (1991).

"Storm Over Biology"

IN HIS 1986 BOOK *Storm over Biology: Essays on Science,*

[10] The title speaks for itself.

Sentiment and Public Policy,[11] the late Bernard D. Davis, a respected molecular biologist at Harvard Medical School, protested the spread of Marxist disinformation concerning human biology, and charged *Science for the People* with being a Marxist-leaning organization which aimed "to destroy the field of human behavioral genetics" (1986). The names of Richard Lewontin, Stephen J. Gould and Leon J. Kamin, in particular, feature prominently in this connection. Lewontin's 1984 book, *Not in Our Genes*, co-authored with Kamin and Britain's arch-antihereditarian Stephen Rose, was an outright attempt to deny the importance of heredity. A self-described Marxist, Lewontin was a prime force behind *Science for the People*, although both he and Kamin have been forced to concede that there is a genetic component to intelligence. Stephen J. Gould – who according to *Skeptic* has denied being a Marxist but says that "he learned Marxism on his father's knee"[12] – similarly advances the cause of biological egalitarianism despite the fact that as a scholar he seesm to be fully cognisant of ongoing research. He appears to have an ideological commitment to what he believes the facts *ought* to be rather than to what they are. Indeed, his book *The Mismeasure of Man* (1981) has created much confusion in the minds of its readers concerning the validity of evidence for the heritability of intelligence.

The determined efforts of Kamin, Gould, Lewontin and Rose to argue that individual human differences are not significantly influenced by genetic forces leads one to believe that they react fearfully against evidence to the contrary because it threatens the ideology around which their personalities have evolved.

Because of media bias favoring the fantasy of the biological uniformity of all mankind, there is little supportive publicity for those like Bernard Davis. An outstanding scholar

---

[11] Bernard Davis, *Storm Over Biology: Essays on Science, Sentiment and Public Policy*. 1986. Buffalo: Prometheus.

[12] *Skeptic*, Vol. 3 No. 4 (1946), p. 26.

with an unimpeachable scientific reputation, Davis wrote many articles documenting leftist distortions of scientific knowledge about heredity. He showed how these distortions were promoting irrational public policies based on the false premise of biological egalitarianism – and on the idea that contemporary *Homo sapiens* were somehow exempt from the laws of genetics, and thus were completely malleable and readily subject to social manipulation. Many of these articles were republished in his *Storm over Biology*, which received only adverse attention from the media, and on his death the initial *New York Times* obituary completely ignored his extraordinary scientific achievements and concentrated on representing him to the public as a holder of fringe views. Only under pressure from Nobelist Arthur Kornberg did the *Times* eventually publish a more objective account of Davis's achievements and scientific status. (Holden, 1994).

VIII
TWIN STUDIES, BLOOD GROUP RESEARCH AND MODERN GENETICS

BUT DESPITE THE post-Boasian malaise which continued to spread throughout the social sciences as a result of the post-World War II victory of egalitarianism, a number of outstanding scientists nevertheless continued to do key research into the role of heredity in shaping human personality and abilities, even though the media gave them little publicity. Typical of these was the world-renowned botanist and geneticist, C. D. Darlington, author of numerous texts including *Genetics and Man* (1964) and *The Evolution of Man and Society* (1969), who warned that selection changed the character of the world's population in every generation, and that "[t]he nation which takes most serious thought for its own genetic future is, therefore, most likely to have a future".[1] In 1968, Frederick Osborn argued for the increasing need for eugenics to eliminate harmful genes from the human gene pool in *The Future of Human Heredity: An Introduction to Eugenics in Modern Society*. Similarly John R. Baker, emeritus reader in cytology at Oxford University, compiled a landmark study entitled *Race* (1974). This catalogued human subspecies and populations around the world and throughout recorded history, and linked these to the different levels of civilization attained, questioning whether the future of humanity was endangered by dysgenic modern trends.

The famed Linus Pauling also warned about the dysgenic threat facing contemporary humankind:

---

[1] From the conclusion of his article "The Control of Evolution in Man," published in *The Eugenics Review*, Vol. 50, No. 3, October 1958.

> I am unhappy that the pool of human germ plasm, which determines the nature of the human race, is deteriorating. The collection of molecules of deoxyribonucleic acid that will make the next generation of human beings what it will be is not so good as that which determined our character; there are more bad molecules in the collection. The defective genes are now not being eliminated from the pool of human germ plasm so rapidly as in the past, because we have made medical progress and have developed feelings of compassion such as to make it possible for us to permit the individuals who carry the bad genes to have more progeny than in the past. Moreover, defective genes are being manufactured at a greater rate than in the past, because there are new mutagenic agents operating in the world today.[2]

Ernst Mayr, the renowned biologist, in 1963 stressed the link between heritability and intelligence, and the fear that the Western world might not be passing on its best genes:

> Nor can there be any question that intelligence has in part a genetic basis, and it is quite immaterial for our argument whether the heritability of intelligence is 25 percent or 75 percent. Finally, there is abundant statistical evidence that in most communities those people whose professions require high intelligence produce on the average smaller families and at a later age than do people like unskilled laborers, whose professions do not make such requirements ... The weight of the available evidence fully supports Huxley's conclusion that those who are intellectually best endowed contribute less to the gene pool of the next generation than do the average and most of the less than average.[3]

But it was Berkeley psychologist Arthur R. Jensen who

---

[2] Linus Pauling, "Molecular Disease," *American Journal of Orthopsychiatry*, Vol XXIX, No. 4, 1959, p. 684.

[3] Ernst Mayr, *Animal Species and Evolution*, 1963, Cambridge: Harvard University Press. p. 659.

succeeded in bringing the role of heredity to the attention of social scientists with his paper entitled "How Much Can We Boost IQ and Scholastic Achievement?" This appeared in the *Harvard Educational Review* in 1969[4] and created shock waves amongst the ensconced egalitarians. His subsequent books, notably *Educability and Group Differences* (1973) and *Bias in Mental Testing* (1980), explained the importance of genetics in determining human intelligence in even greater detail. Further attention was also drawn to the subject of human heredity by University of London psychologist Hans Eysenck in England with his 1971 book *Race, Intelligence and Education*, and by C. H. Waddington's *The Evolution of an Evolutionist* (1975) and *The Man-Made Future* (1978). In 1982, Raymond B. Cattell demonstrated beyond reasonable doubt the high degree of heritability of fluid intelligence in his book *The Inheritance of Personality and Ability* (1982).

This new awareness was reinforced by E. O. Wilson's *Sociobiology: The New Synthesis* (1978), which accepted an evolutionary approach to the study of the social sciences. Another notable scholar who sought to emphasize the need to recognize the importance of the heritability of human differences was the Nobel Prize-winning physicist and co-inventor of the transistor, William B. Shockley, many of whose writings on the subject have been reproduced in *Shockley on Eugenics and Race* (Pearson, 1992).[5] In the U.S.A., Richard J. Herrnstein's famous Atlantic article entitled *I.Q.* (1971), and his subsequent book *IQ in the Meritocracy* (1973), stressed the increasing importance of intelligence in today's complex civilization. Herrnstein explained that as intelligence is largely hereditary, any innate deficiency in that respect would bar the affected individuals from career success, and that, as technology advanced, "biological stratification" would

---

[4] HER Vol 39 (1): pp. 1-123.

[5] Roger Pearson, *Shockley on Eugenics and Race*, 1993, Scott-Townsend Publishers, Box 34070, NW Washington DC 20043. Introduction by Arthur R. Jensen.

become increasingly obvious. Herrnstein joined James Q. Wilson in co-authoring *Crime and Human Nature* (Wilson, 1985) which raised the possibility of a genetic basis for sociopathic defects in personality. Then, shortly before his death, he co-authored (with Charles Murray of the American Enterprise Institute) *The Bell Curve: Intelligence and Class Structure in American Life* (1994).

The Bell Curve had a major impact on the public and scholastic world alike – despite what appeared to be an organized and devastatingly adverse media reaction by committed egalitarians who saw their moral base crumbling beneath them. Herrnstein and Murray warned that today's massive social problems cannot be solved without recognizing the role of heredity in determining the limits of human behavior, including its impact on social issues such as crime, illegitimacy, welfare dependence, productivity, and social stratification. Differential birthrates, they pointed out, impact on national intelligence, and the failure of school integration and other experiments intended to raise black educational performance can only be understood fully by taking genetic as well as environmental factors into account.

Lay writers also are beginning to explore these subjects and make them understandable to the general public. Daniel Seligman, senior editor of *Fortune,* has published a highly readable book called *A Question of Intelligence: the IQ debate in America* (1992) covering what scientists know and do not know about intelligence and why intelligence is so important to human beings. Writer Peter Brimelow (1993, 86) of *Forbes* magazine said that Seligman "defies one of the most powerful contemporary taboos" by reporting the importance of intelligence and its heritability. Charles Murray (1992, 63) stated that the Seligman book revealed "labyrinthine ramifications for our culture and our polity."

Rushton's findings are even more striking. He has claimed to find that East Asians and Africans score at opposite ends of a continuum, with Europeans intermediate, in more than 60 anatomical and social dimensions including brain size, crime rates, family structure, personality, and sexual behavior. He has shown conclusively that amongst living hominids brain size is correlated with IQ test scores (by using magnetic resonance imaging techniques to contrast, *in vivo*, 3-dimensional pictures of the brain) and that East Asians and Europeans average about 100 cubic c.c. (5 cubic inches) more brain matter than Africans. From a stratified random sample of 6,325 U.S. Army personnel, Rushton (1992)[6] calculated average cranial capacities for Asians, Whites and Blacks and found that this corresponded to the racial ordering in IQ. Much of his work was summarized in his book *Race, Evolution, and Behavior* (1995).[7]

These and other scientists have now produced a wealth of evidence demonstrating beyond reasonable doubt the gross absurdity of biological egalitarianism – and of the many environmentally-based social programs that are not only economically wasteful but have seriously dysgenic implications. The even more dramatic possibility is that with the present demographic trends the percentage of the intellectually well-endowed members of the total world population is likely to undergo a dramatic decline relative to those who are less-well endowed in this respect.

Twin Studies Reveal the Role of Heredity in Shaping Human Personality

In the field of psychology, twin studies in particular have produced a compilation of data which demolishes the fanciful theories of of biological egalitarianism, providing quantitative evidence of the importance of heredity in human behavior. The Minnesota Twin Study has shown that the evidence for genetic influence on human cognitive abilities, as

---

[6] J. Philippe Rushton, "Cranial Capacity Related to Sex, Rank and Race in a Stratified Random Sample of 6,325 U.S. Military Personnel, *Intelligence*, 1992, 16:401-413.

[7] J. Philippe Rushton, *Race, Evolution, and Behavior*, 1995. New Brunswick, NJ: Transaction Publishers.

well as on general aspects of personality, is overwhelming. Tom Bouchard, director of the project, has reported that the degree of similarity between twins reared apart in different environments is much greater than between fraternal twins raised in the same family. The results of personality tests also reveal striking similarities in cases such as that of the personality profiles of a set of identical twins, one of whom had been raised in Trinidad by a Jewish father and the other in Germany by a Catholic mother. Medical life histories also provide extremely powerful evidence of the importance of genetic influences on behavior. Brain patterns show a higher degree of similarity between identical twins reared apart than between fraternal twins reared together, and medical and dental case histories are closely parallel even where identical twins have been reared under quite different circumstances.[8]

Thus, identical twins separated by adoption after birth and reared in disparate environments reveal strong similarities which environmentalists cannot explain away. In fact, the similarity in personality and intellectual ability of identical twins reared apart is little different from the similarity shown by identical twins reared together.[9] These traits include not only shared psychological and vocational interests, but even psychological proclivities such as conservatism, sociability, self-control, flexibility and religiosity as measured by academically accepted standards of assessment. This suggests that the members of distinct breeding populations (i.e. gene pools) must be expected to inherit diverse proclivities for different kinds of cultural behavior, and that patterns of life style and even the character of developed civilizations might well be shaped by inherited personality biases in addition to the

---

[8] See Nancy Segal, "The Nature vs. Nurture Laboratory" in *Research*, July/August 1984.

[9] See amongst other literature "Personality Similarity in Twins Reared Apart and Together," By A. Tellegen, D.T. Lykken, T.J. Bouchard, K.J. Wilcox, N.L Segal and S. Rich, in *The Journal of Personality and Social Psychology*, Vol. 54, No 6 (1988).

obvious limitations imposed by disparate levels of intelligence.[10] As a result, committed egalitarians such as Leon Kamin strive to denigrate the findings of twin research such as that conducted at the University of Minnesota by suggesting that twins reared apart might "have particularly strong motives to downplay previous contacts and to exaggerate similarities."[11] One suspects that these critics are pushing their arguments beyond the limits of their own logic, and that they know in their own minds that such twins could hardly fabricate similar brain waves, blood pressures, heart rates, cardiovascular functioning, pulmonary functioning, electrodermal response amplitude, occurrence of dental cavities, and reaction time to various novel tasks – and that such physiological similarities as these could not be modified by occasional or even frequent meetings, had these in fact taken place.

Blood Groups and DNA

The days when pro-egalitarian propagandists, totally disregarding what was known even in their time, could persuasively argue that individual and group differences were "only skin deep" are gone forever. The days when it was claimed that the technique of blood transfusions, first widely practiced in World War I, demonstrated the egalitarian unity of all members of the human race are also gone, for it is now known that while some of the major blood groups are shared by all races, they are also shared with primates and various other mammals. During World War II great emphasis was laid by egalitarian propagandists upon the fact that the ABO

---

[10] There is a mountain of literature dealing with twin studies, but a synopsis of the racial implications may be found in "Comment on the Minnesota Transracial Adoption Study" by Michael Levin, published in *Intelligence*, Vol 19, 13-20 (1994).

[11] An objection that the similarities between twins might be due to contact between twins prior to testing was raised by the dedicated biological egalitarianist T. J. Rose, in respect of a sample of Finnish twins, but this possibility has been fully considered by T. J. Bouchard and Matthew McGue (1990).

blood group system (the identification of which is so essential for successful blood transfusion) characterized all human races, without realizing that these basic blood group categories were not only common to all human populations, but were also to be found in other primate and mammalian species. It is time that we stopped promoting the specious concept of the uniformity of humankind, and on the one hand recognized the unity of humankind with the primates and all other living organisms, while on the other hand appreciating the immense significance of what have hitherto been regarded as relatively small differences between human individuals and between human breeding populations.

John H. Beckstrom emphasizes the evolutionary importance of kinship and "familial genes", commenting that:

> All humans alive today share, in common, the vast majority of their genes. But over and above those genes that prevail in the entire species, additional gene commonality exists between close relatives because they received their genes from close ancestors whom they have in common .... [–] 'familial' genes." (1993 p.15)

As additional blood group research was conducted, it was found that there were many more systems other than the ABO. Thus, it is now common knowledge that not only does the frequency of distribution of blood groups vary from one population to another, but that many populations, especially genetically isolated populations such as may be found in remote parts of the Arctic, Siberia, South America and Central Africa, reveal rare blood group polymorphisms which are specific to their own populations and reflect the uniqueness of their evolutionary histories. Distinctive genotypes are inevitable among groups which have experienced long periods of separate genetic evolution.

That it was possible to classify races by blood groups was early revealed by William C. Boyd in his *Genetics and the Races of Man: An Introduction to Modern Physical*

Anthropology[12], and by A. E. Mourant in his *The Distribution of the Human Blood Groups.*[13] This latter contained a foreword by the renowned physical anthropologist, H. J. Fleure, who declared that "the study of individual [blood group] characteristics and of their relations with one another, especially in their geographical distribution among percentages of different populations, is the most hopeful road towards understanding human diversity."

Mourant found that blood group patterns not only distinguish the major races of the world quite sharply, but also that minor differences could even be used to classify the different peoples of Europe, with Sardinia and Sicily differing markedly from northern and central Europe. A dramatic divide separates Europe and sub-Saharan Africa. Indonesia and Australia are also very different from Asia, while American Indians have been sufficiently isolated in their respective territories to have evolved quite varied blood patterns.

Boyd and Mourant's pioneering studies formed the foundation for much more specialized research by later scholars, and courts of law today accept blood group evidence in legal disputes regarding paternity, It is also widely accepted that the variations in the frequency of specific blood groups that distinguish genetically separate human populations from each other carry within them a coded guide to the racial history of each breeding population. Relying on this, several scholars have successfully used blood group analysis to throw light on the demographic history of the Jewish diaspora,[14] utilizing blood group patterns to determine not only the extent to which the various Jewish communities had interbred with the local populations amongst whom they had resided over the centuries, but also to gain clues as to which

---

[12] Published by Little Brown, Boston. 1950

[13] Charles C. Thomas, Springfield, Illinois, 1954

[14] E.g. A. E. Mourant, A. C. Kopec and K. Domaniewska-Sobczak *The Genetics of the Jews,* 1978

populations they had had contact with during this time. As information accumulates about the blood groups of more remote populations a clear world picture is beginning to evolve, which is being fortified by even more dramatic advances in the study of human molecular genetics.

IX
DNA ANALYSIS AND THE EVOLUTIONARY HISTORY OF MAN

THE BASIC UNITS of inheritance are commonly known as genes, even though this term is now understood to be purely symbolic, representing as it does strands of deoxyribonucleic acid (DNA) bearing a complex array of nucleotides arranged in particular sequences. The genetic mechanism is responsible for the production of protein molecules composed of aminoacids. Structural alterations in the strands of DNA are termed mutations. Sickle-cell anemia, for example, is a genetic disease that can result from the substitution of just one aminoacid out of about 140 aminoacids that make up each of the four polypeptides that comprise the protein hemoglobin. Genetic research is aimed at understanding this mechanisn so as to achieve the eugenic goal of being able to correct deleterious mutations that handicap living organisms. Most mutations create an imbalance in the complex biological structure of the organism, thereby causing it to function less efficiently (i.e. less competitively), or even to malfunction altogether (to die or fail to reproduce).

The ability to analyze and classify human DNA has advanced rapidly in recent years, and DNA testing is supplementing if not supplanting blood group analysis for forensic purposes. An individual's DNA is as unique as his fingerprints, yet it also reveals evidence as to which other individuals he has genetic or racial ties. A single hair, a sliver of skin, even a dandruff flake can reveal the individual's genetic identity, and some artists are now incorporating a sample of their DNA into the signatures on their paintings in order to facilitate the detection of forgeries.

As with blood groups, attempts to minimize the

importance of differences in the DNA of living human populations are nothing more than a red herring. The vast majority of human nucleotide bases, out of an estimated total of some three billion, are shared with other non-human life forms: it has been estimated that of these three billion only between two to ten million differ from person to person. According to one estimation of genetic relationship, human DNA differs from Chimpanzee DNA by no more than 1.6 per cent overall.[1] Yet this 1.6 per cent is of enormous importance.

DNA profiling has already begun to be used to assist historians in determining the racial history of populations by revealing their genetic components. L. L. Cavalli-Sforza has carried out extensive research into the genetic history of diverse human populations (1988, 1993), and with co-authors Paolo Menozzi and Alberto Piazzi, has embarked, in *The History and Geography of Human Genes* (1994), on a pioneer odyssey into the historical and genetic relationships that demarcate living populations of humankind throughout the world.

Although based on a mere 120 alleles drawn from only 42 living populations, this study points the way to further research which in time could untangle the complex web of historic migration and genetic admixture with a degree of precision which would have been impossible with the more primitive techniques known to physical anthropologists just fifty years ago. Summarizing the worldwide data in a singular 42 by 41 matrix of genetic distances, Cavalli-Sforza and his co-authors find that the English differ from the Danes, Germans

---

[1] More cautiously, Japanese anthropologists Y. Watanabe and S. Uedo of the University of Tokyo have concluded that humans and chimpanzees share at least 95% of the nucleotide sequences of their DNA. They warn that the full significance of genome dissimilarity between species (and by implication between subspecies) is as yet but little understood and that even minor chromosomal disparities can make a major difference in the morphology and behavioral abilities of an organism.

and French by a mere 21-25 points of genetic distance, whereas they differ from North American Indians by 947 points, from the African Bantu by 2288, and from Mbuti Pygmies by 2373. An erstwhile dedicated environmentalist, in respect of race and IQ, Cavalli-Sforza genuflects in a few mandatory paragraphs to the crusade against "racism," but concludes that the collected data he presents show that "North Eurasians," – a term he uses to group Europeans, Mongols, Japanese, Koreans, and Northern Chinese – are more closely related to each other than to Africans and Australoids.

In their determination to ignore the traditional anthropometric terms which classify living populations along racial lines, Cavalli-Sforza and his co-authors have chosen to use purely geographical or national terminology, which has little to do with genetic realities in many parts of the modern world. Yet even though they are working with non-adaptive (i.e. non-racial) genetic markers which were chosen because they had no known evolutionary implications, the racial patterns that are revealed coincide closely with the conclusions of traditional anthropologists. The argument about whether or not "races" exist is purely a matter of semantics – of what exactly the debaters understand by the term "race". Distinct breeding populations, each its own gene pool, must have existed since soon after sexual reproduction first evolved: otherwise, there could have been no speciation and no evolution. If we apply the term "race" in a broad sense to refer to any group of people who, as a result of mutation, genetic drift and evolutionary selection, have evolved recognizably distinct sets of genetic characteristics, the term presents no scientific difficulties whatsoever. In short, studies of genetic markers group living populations of *Homo sapiens* into categories that are essentially parallel to those determined by the older-school physical anthropologists represented by Carleton Coon.

The problem arises solely from contemporary "political correctness", or political convenience. The demographic realities of the world, and the fracturing of national boundaries under burgeoning migratory pressures, are converting the Western nations into multiracial states (something that is *not* happening in the more prolific countries of Asia and Africa). This is a situation which is clearly fraught with the potential for conflict. Any rise in consciousness of biological identity among the indigenous Caucasoid populations of the West – and resultant demands for the repatriation of biologically alien immigrants before they become a majority – is regarded as politically tendentious. There is consequently a drive to condemn "racism" and "xenophobia" as dangerous reactions, likely to touch off major tensions if permitted to grow in strength, and it has become politically expedient to suppress nationalist feelings and reduce popular awareness of genetic differences by lending broad support to the myth of biological egalitarianism.

But despite the fact that human populations tend to grade into one another, as physical anthropologists[2] had recognized from an early date, the current use of the term "race" to identify specific populations – as distinct from its historically loose usage to refer to *any* descent group – complies with what Max Weber called an "ideal type": a stylization of concrete reality which may not fully subsume reality but which is a useful, often indispensable, instrument for conceptualizing reality. As Max Weber explained, such intellectualized categorizations are not necessarily definitive or exclusive. They are valuable mental tools – like "hot" and "cold" – in that they approximate reality: they conceptualize extremes of variation in the same way that mountain ranges define a topographical map.[3]

---

[2] See in particular Carleton Coon, *The Origin of Races*, 1972, and *The Living Races of Man*, 1965, both Alfred A. Knopf, also John R. Baker, *Race*, 1974, Oxford University Press, for a more detailed explanation of the anthropological utility of the term "race".

[3] For a discussion of Weber's use of the "ideal type" as an aid to the comprehension of complex reality, and its indispensable utility to

Mitochondrial DNA

Interestingly, Cavalli-Sforza's studies support the findings of mitochondrial DNA (mtDNA) research which suggest that at one time the ancestors of all living hominids originated in Africa and that the major racial differences arose as a result of an early migration of certain hominids from Northeastern Africa into Eurasia. These migrating groups were subjected to subsequent rapid evolution in colder more northerly zones, while those that remained in Africa were not affected by the selective forces associated with the harsh new environment and so experienced less dramatic evolutionary change. This is reinforced by DNA evidence, which reveals a primary divide between the contemporary populations of sub-Saharan Africa, and those that evolved their present, highly distinctive characteristics in Eurasia in sharply different, heavily challenging, environments.[4]

DNA theorists often differ from paleoanthropologists (who study actual fossils) in assessing the antiquity of race differences. Carleton Coon and modern paleoanthropologists such as Milford Wolpoff[5] point to surviving skeletal evidence which reveals that certain present-day racial characteristics, such as the Mongolian shovel-shaped incisor, were already present in pre-*sapiens* populations, indicating either separate regional evolution of contemporary populations from pre-*sapiens* ancestors or significant interbreeding of intruding *Homo sapiens* populations with local pre-*sapiens* populations. Mitochondrial DNA enthusiasts insist that early *Homo sapiens* populations maintained total racial separation, and did not crossbreed with females belonging to other subspecies of

---

epistemology, see Max Weber's *Sociology of the Intellectuals*, by Ahmad Safri, 1992, Oxford University Press, pp.11-22.

[4] Populations which separated from a common ancestral population more recently will show less variation in their mtDNA.

[5] See Milford Wolpoff, "Multiregional Evolution: The Fossil Alternative to Eden" in *Behavioral and Biological Perspectives on the Origins of Modern Humans*, edited by F.H. Smith and F. Spencer. 1989, New York: Alan R. Liss.

Homo when expropriating the territories of the latter. Paleoanthropologists, by contrast, suggest that rapidly evolving subspecies tended to expand into territories occupied by more primitive subspecies and probably interbred with these.[6]

However, DNA evidence pointing to the African continent as being the original point of dispersal from which the ancestors of other hominids departed to colonize other parts of the world, fits closely in all respects, other than the date of hominid dispersal from Africa, with what paleoanthropologists have deduced from the fossil record. Clearly early primates like those of today were located in tropical or near tropical climates, and rapid evolution took place as a result of kladogenesis among those hominids who managed to move out of those areas into more temperate zones. Those which remained in the tropical regions would not have been under the same selective pressures to evolve differently. Thus the African negroes, and the negritoes descended from early migrants who never experienced the Ice Ages and settled in the tropical rain forests of Southeastern Asia, may consequently be expected to resemble their earlier forebears more closely than the members of races which underwent severe selection in the colder climates of Northern Eurasia.

Mitochondrial DNA is only one of some 40,000 gene linkages present in humans. The work that has been done to date covers only a very small portion of the human genetic heritage, and very much more data will come to light as

---

[6] This brings to mind the old dispute about the evolutionary history of the Neanderthals (today regarded as *sapiens* because their distinguishing skeletal characteristics can be matched among various extant living populations), and their subsequent displacement by Cro-Magnons. Some paleoanthropologists believe that interbreeding may have taken place in some areas between purely *Homo sapiens sapiens* Cro-Magnon and Neanderthals, just as Carleton Coon speculated that the descendants of Java Man might have been raised to *Homo sapiens sapiens* status by interbreeding with incoming *H.s.s.* invaders, to create the modern Australoids.

research continues. While some recent research now suggests that a common ancestral hominid population existed as late as 130,000 years ago, research by Dorit, Akashi and Gilbert[7] on the Y chromosome, which is inherited through males (as opposed to mtDNA which is inherited only through females), suggests a possible date of 270,000 years ago for a common ancestral hominid population. However, their calculations indicate a chance that the correct date could be as far back as 800,000 years ago. If an earlier date proves to be correct, then the contradictory data presented by fossil evidence, such as the shovel-shaped incisor, which militates against the idea that all living peoples are descended from a single parent population that existed only 130,000 years ago, would no longer be problematic. The shovel-shaped incisor, which was present among the hominid inhabitants of Northern China some 350,000 years ago, is still common among contemporary North Chinese, and is found, less frequently, among more distantly related Mongoloid populations, such as the Vietnamese, the Inuit, and even some American Indians.

It is quite probable that further research will modify the present estimates of the rate of DNA mutation in a way that may bring this research into line with the paleoanthropological fossil evidence.

Genetic Markers Confirm Traditional Race Classification

The traditional classification of the world's peoples by physical anthropologists, based primarily on anthropometric and somatic data, into identifiable races can only be understood by those who realize that human races are *not* to be thought of as clearcut and mutually exclusive categories. The classification into human races is a conceptualization which distinguishes different, recognizable clinal peaks, with intermediate populations filling the interstitial spaces just as valleys separate mountain peaks. As

---

[7] Published in *Science*, 268: 1183-1185. May 26, 1995.

a Weberian technique, racial terminology has proved itself since its first inception early in the mists of history to be a useful working tool, and the latest biological research reinforces the reality of these clinal peaks.

In some situations the term "race" is therefore more useful than the term "subspecies," with which it is today often equated. The term subspecies tends to connote a static biological group identified at a particular moment in time. But the biological reality is the phylogenetic continuum, which exists in time and can evolve from one species into an entirely new species in the course of ongoing evolution. The term race by no means precludes the concept of genetic admixture, of ongoing evolution, of eugenic transformation, or even of dysgenic transformation; and terms associated with the concept of race emphasize the changing character of phylogenetic continuua in time.[8]

Since his 1988 article entitled "Reconstruction of human evolutionary biology: Bringing together genetic, archaeological and linguistic data" in which he used traditional anthropological race terms such as Caucasoid, Cavalli-Sforza has repeatedly asserted that his study of the history of human genes is not "racist" and that human beings are "all the same beneath the skin"[9] – as though human blood does not flow beneath the skin, as though bones do not differ in density and as if all hominid brains are of the same size and shape and equally developed in all areas and functions. However, N. Saitou,[10] amongst others, has identified six major racial

---

[8] The reality of intergenerational genetic change in human breeding populations was strongly recognized by the early eugenicists – witness the historic use of such terms as "racial hygiene", "racial improvement," "race decay" and "race suicide." All refer to the fact that life forms are not static, but constantly undergo change, for better *or* for worse.

[9] Reference his remarks quoted in *The Washington Post*, ...

[10] "A Genetic Affinity Analysis of Human Populations, in *Human Evolution* (1995) 10:1, 17-33. Saitou is attached to the Laboratory of Evolutionary Genetics at the National Institute of Genetics in Mishima, Japan.

groups dating from the beginning of the Holocene, which correspond to traditional racial groupings determined by anthropometrists. Although the picture has become blurred due to increasing geographical mobility, Saitou believes that "the global movement of ancient human populations until the end of the Pleistocene can be traced even if we use the data from present-day populations, because human movements on the globe seem to have accelerated only after the Neolithic Revolution started, namely in the last 10,000 years." He adds, "These topographical patterns of the genetic affinity tree ... closely resemble the geographical constellation of the six geographical areas: Africa, West Eurasia, East Eurasia, North America, South America, and Sahul land."[11]

Two additional racial groups, the Negritoes and Capoids, are barely discussed by Saitou, since little biochemical research has been undertaken on these now scarce and remote peoples. However, the close parallel between the data revealed by biochemical markers and the traditional pattern of anthropological race classification is strikingly apparent.[12]

The hominid populations that remained longest in the relatively constant environment of the tropical world, and were therefore less exposed to new selective pressures requiring morphogenetic adaptation, were the Negroids of sub-Saharan Africa. In this context the term includes the paleo-Negroids (Sudanids and Bantu), the Nilotids and the Pygmies,[13] all labelled by Saitou simply as African. Biochemical analysis of non-adaptive, or less-adaptive, genetic markers reveals a wider diversity of such markers among the

---

[11] Saitou's Sahul land refers to the Sahul shelf, which existed until the end of the last glacial period, when it became divided into Australia and New Guinea.

[12] For clarification concerning the traditional anthropological terms used in this section, and for an overview of the evolutionary interrelationship of physical and cultural anthropology, see Pearson (1974) and Pearson (1985).

[13] These terms are taken from the classification adopted by Hans Jürgens, *The Face of Africa*, 1981. Washington DC: The Cliveden Press

present-day peoples of Africa than amongst the peoples of other continents. This is because the markers that selected for study are those which are relatively "neutral" from an evolutionary point of view. Since it is assumed that mutations occur at a more or less regular rate[14] populations which show little difference in the non-adaptive genetic markers may be assumed to have separated a shorter time ago than those which show a greater degree of variation, i.e. they are more closely related. This is only true of genetic markers which are neutral from an evolutionary point of view – those which have little or no selective survival value. Racially-important genetic characteristics, which change in response to evolutionary challenges, cannot be used to judge the passage of time, since they will tend to be similar not only where populations have only recently separated but also where parallel or convergent evolution has occurred under similar selective conditions.

Thus, the peoples of Eurasia, America and of Australasia can be shown by DNA research to have long been separated from the indigenous populations of Africa, who have remained in essentially the same environment that originally nurtured them. On the other hand, African populations show a wide disparity of non-adaptive genetic markers, but share many similarities in their adaptive genetic make-up due to having remained in the same selective environment. Thus the paleo-Negroids of the tropical forests of central and West Africa can be shown to be closely related

---

[14] This principle may yet be subject to further refinement, as other dating methods, such as carbon 14, have been. Mutation rates may well vary at different periods of time, possibly due to radiation from outer spacc, but this would presumably affect most populations and so would not seriously impair the utility of comparing the number of mutations in different populations to determine how long they have existed as separate breeding isolates. A more serious problem is that a large population with much fissioning will show a greater number of mutations (Harpending and Retherford, *Current Anthropology,* 1994). Notwithstanding this, most specialists, including Saitou, work on the assumption that mutations provide a reliable biological clock.

to the Bantu, who in historical times spread southwards and eastwards across Africa, as well as to the Nilo-Saharans of the Upper Nile and southern fringes of the Saharan desert. More distantly related to the Negroes are the Pygmies of the rain forests, and less closely related still are the San or Bushmen of southern Africa, who are believed to have once covered most of the unforested savannah grasslands of Africa. By contrast, the Ethiopians and Somalis are derived from a Caucasoid population which became admixed with Negroids, and possibly also with early Bushmen elements. But again, one must not discount the possible survival of very ancient genetic differences, reflecting genetic clines between more extreme centers of biological disparity.

While it is now hypothesized that the ancestors of all living peoples originally came out of Africa, it is more than probable that those who moved into Asia would have lived in the Horn of Africa, or the region of the Upper Nile, where paleoanthropological evidence indicates the savannah highlands had since early times been parent to more rapidly evolving hominids. It is entirely possible, therefore, that the Ethiopid and Somali peoples are not wholly the product of Caucasoid immigration into Africa from Asia – as to some extent they must surely be – but may well be in part directly ancestral to the peoples who exported their surplus population to find new territories in the more northerly parts of the Nile valley and in the adjacent Eurasian land mass.

Cavalli-Sforza and his co-authors agree that the sub-Saharan African peoples are markedly different from the other peoples of the world, and most anthropologists believe that they probably remained closest to the ancestral hominid population, presumably because they remained in the same environmental niche and were not subjected to the disparate selective pressures which resulted in further evolution among those who moved into different parts of the world, and became subject to fierce challenges for survival.

The inhabitants of Western Eurasia, identified by traditional anthropometrists as Caucasoids, are closely linked to the majority element in South Eurasia, otherwise known as the Indian subcontinent. The term Caucasoid was originally chosen because of the more or less central location of the Caucasus mountains in this geographical area. Saitou (1995) labels this group with the simple geographical term, West Eurasian. Within this aggregate of recognizably distinctive peoples the more extreme clinal peak is represented by the relatively homogeneous peoples of Northern Europe. They, more than any other Caucasoid people, represent the clinal extreme of the Caucasoid populations, being characteristically more different from all other extreme hominid variants currently inhabiting the earth. This is due not only to the nature of the environment in which they evolved but also to the fact that – with the exception of the Lapps of Northern Scandinavia and the Samoyeds who live adjacent to the Nordic "Great Russians" – in prehistoric and even in historic times they were more isolated than other Caucasoids from contact with the non-Caucasoid races of Africa, Eastern and Southern Asia.

The analysis by Cavalli-Sforza and his co-authors of biological data relating to 42 ethnic groups from around the world, and supporting blood group analysis, shows clearly the close relationship of the North European peoples (excluding the Lapps and Samoyeds), with a slight but identifiable variation from the epicenter in North Germany, Southern Scandinavia, the British Isles, Northern France and the Low Countries, as the analysis moves eastwards through Poland into Northern Russia. Closely related to this North European population are the peoples of Southern Europe, with Sardinians and Basques varying recognizably from the population of the genetic epicenter. Modern Greeks differ even more, coming closer to the still-Caucasoid Iranians. The population of the Northern part of the Indian subcontinent varies from the Caucasoid epicenter as also does the population of Southwest Asia (the Middle East). The peoples of North Africa, such as the Berbers, likewise differ from the Caucasoid extreme, but still fall more clearly into that category than into any other racial grouping. The population of Southern India is also placed within the Caucasoid macro-race, but unlike the population of the northern part of the subcontinent, who are believed to be largely descended from immigrant Indo-Europeans, they have Dravidian origins, originally linked to the Southwest Asians, but appear to have absorbed genes from earlier populations which were closer to the Australoid clinal peak, with occasional evidence of Negrito influence. This would make sense if the forebears of the Negritoes and the Australoids had moved to their present locations by migration through Southern Asia.

The population of Northern China appears to mark another clinal center, which Saitou has labelled East Eurasian but was known to traditional anthropometrists as the Mongoloid macro-race. Cavalli-Sforza *et al* note that DNA and blood group evidence indicate that the Japanese, Koreans and Mongols are genetically closer to the North Chinese and the Tibetans than to other racial groups, and Northern China and Mongolia have been classified by physical anthropologists as the clinal center of what has been called the Mongoloid race. The Samoyeds who still occupy territories in Europe north of the Great Russians are genetically close to the center of this category, and the Lapps may represent a related people who because of their position in Northern Scandinavia have through the millennia received a major genetic contribution from the Nordic Caucasoids, and so are today closer to the Caucasoid medium, although still markedly separate from the vast majority who are grouped within this terminological category.

Popular usage of the term Mongoloid is extremely loose, because in historic times the Mongols of Northern China have succeeded in multiplying and invading adjacent territories to the south and even into the Pacific as far as the Philippines and the Southeast Asian archipelago. Four thousandyears ago South China is believed to have contained peoples of Australoid character, while the rain forests of Southeast Asia were inhabited by Negritoes, a diminutive people who outwardly resemble African Negroes. In any early migration out of Africa it is possible that those hominids who followed a South Asian tract and settled in Asian tropical forests would have retained, and been selected for, those same protective characteristics such as heavy pigmentation and peppercorn hair that mark out the inhabitants of the tropical forests of sub-Saharan Africa. Their blood group patterns have evolved differently, however, and so the Negritoes, although rare today, mark another clinal peak of biological divergence.

Another major racial group which traditional anthropological studies have indicated was once widely spread over South and Southeast Asia, from India and South China down to Australia and Tasmania, are the Australoids, identified by Saitou as the Sahulians (referring to the indigenous populations of which still survive in New Guinea and Australia today). However, as Mongoloids expanded in all directions from Northern China and Mongolia, those that settled in Southeast Asia, in the Philippines and in present-day Indonesia may have become admixed with these earlier populations, and do not represent the Mongoloid type in its more extreme sense. Cavalli-Sforza and his colleagues find that the biochemical markers they have studied show the South Chinese, the Mon Khmer of Cambodia and other parts of Southeast Asia, as well as the Thais of Thailand constitute a separate group within the Mongoloid family. Traditional physical anthropologists would say that this was due to the absorption of Australoid and Negrito genes amongst those who penetrated Southeast Asia. The Indonesian and Philippine peoples are also somewhat segregated biologically, possibly due to having absorbed both Australoid and Negrito genes. Micronesians and Melanesians are distinguished from Mongoloid peoples, possibly reflecting an even larger Negrito component in the case of the Micronesians, and a major

Australoid component in the case of the Melanesians.[15]

Cavalli-Sforza and his colleagues clearly distinguish the aboriginal inhabitants of New Guinea and Australia as a very separate group on their phylogenetic tree. After the major phylogenetic divide that separates the African populations (the Pygmy, Bushman, West African, Bantu and Nilotic peoples on their schemata), all the non-African peoples, according to their calculations, then sub-divide into two major divisions, a Australoid or Sahulian division and a Eurasian and American Indian division – all those peoples classed by traditional anthropologists as Caucasoid, Mongoloid and Amerindian.[16]

Northeast Asia is an unappealing expanse of marginal territory that harbors the remnants of ancient populations which range between the Mongoloid and what appear to be early Caucasoid variants. The Ainu, who were once the prime residents of Northern Japan, Sakhalin, and parts of the Northeast Asian mainland may be of part proto-Caucasoid ancestry, and it is now known that Caucasoids at one time occupied a vast belt along the central Asian steppes, south of the coniferous forest belt, extending as far east as the Altai mountains. The remains of Nordic peoples, who are believed to have spoken a language akin to German and Celtic, have recently been found in Xinjiang (Chinese Turkestan), and graves containing the remains of similar Caucasoids have been found in the Altai mountains. The cold coniferous forests to their north were seemingly always occupied by Mongoloids, and in historic times Mongol hordes moved westwards, raiding into Europe and Southwest Asia, even establishing a Mongol dynasty in India for a period of time. Thus the Turkic and

---

[15] For a convenient handbook listing the diverse human ethnic and racial groups, as well as anthropological terms in general, see Roger Pearson, *Anthropological Glossary*, 1985. Malabar, FL: Krieger Publishing Company.

[16] Negritoes are not detailed in either the survey by Cavalli-Sforza *et al* or that by Saitou, possibly because of the absence of adequate data.

Uralic peoples reflect some Mongoloid characteristics. The last two thousand years have seen genetic waves rolling backwards and forwards across central Asia, confusing any clinal gradients which may originally have existed in that vast area. Since Northeastern Siberia still contains a wide variety of small population isolates which have been but little studied biochemically, the data analyzed by Cavalli-Sforza *et al* and by Saitou do not include these populations. Only a few of the larger Northeast Asian groups, such as the Chukchi, Ainu and Samoyeds, are included in the Cavalli-Sforza study.

And finally, we come to the American Indians. These were generally classed by traditional anthropologists as Mongoloids, and are believed to have entered the Americas from Eastern Eurasia across what we now call the Bering Straits. Their Mongoloid affinities are apparent to physical anthropologists and biochemists alike, although some Caucasoid genetic component seems to be present in certain of the North American Indian peoples, and on the basis of biochemical markers these test closer to Caucasoids than to the Chinese or Japanese. This is not surprising in light of their Eurasian origin, for the population of Central and Northeastern Eurasia contains a complex array of genetic isolates reflecting a variety of Mongoloid and Caucasoid characteristics.

Saitou emphasizes a circum-Pacific relationship between the North American Indians, the South American Indians and the East Eurasians (Mongoloids). Oceania is peopled by Polynesians, Melanesians and Micronesians, with varying admixtures of East Eurasian (Mongoloid) elements, plus Sahulian (Australoid) and Negrito genes in the latter two, respectively. There is also evidence that Caucasoids from the Indian subcontinent early reached Indonesia by sea and contributed genetically not only to the Indonesian population but also to the seagoing Polynesian peoples.

As mentioned earlier, contemporary data suggest that the Mongoloid peoples perform better on IQ tests than the genetically-varied American population today classed as "white", but in reality comprising many of mixed descent who style themselves as white. It may be true, as Richard Lynn and J. Philippe Rushton have suggested,[17] that this modern disparity of intellectual ability can be traced to disparate evolutionary histories that date back as far as the last Ice Age. However, others point to the lack of scientific racial classification where the mixed population of North America and other countries such a Britain is concerned, which includes any person of mixed descent who chooses to designate himself or herself as "white." Furthermore, it has been argued that during the last glacial period selection was as severe in Northwestern Eurasia as in Northeastern Eurasia, but in historical times the peoples of the Far East pursued essentially eugenic patterns of reproduction while those of Europe followed dysgenic cultural practices. In China, the more educated classes (who were likely to be the more intelligent) for some two thousand years maintained concubines and raised large families. Despising military activities, they administered territories and seldom personally engaged in combat. Warfare therefore struck more heavily at the lower classes than at the educated elite. In Japan, it is true that the Samurai enjoyed high social status, yet suffered heavy losses in the pursuit of their martial profession. But they too had concubines, and reproduced prolifically.

In Europe, by contrast, those with intellectual ability often sought social advancement within the Catholic church, where as priests and monks they were required to maintain celibacy and hence seldom passed on their genes. Those competent families which occupied non-celibate leadership positions in society were encouraged by the ethos of the Classical and Medieval period to prove their powers of leadership on the battlefield. This dysgenic tendency remained

---

[17] See Richard Lynn, "The Evolution of Racial Differences in Intelligence" in *The Mankind Quarterly*, Vol. XXXII Numbers 1 & 2, Fall/Winter 1991, pages 109-132 and J. Philippe Rushton, *Race, Evolution and Behavior* (1995).

true during the two World Wars fought in this century – especially World War I, when the average length of life of a young officer in any of the major battles was said to be 30 days, and for the pilots of warplanes only half that. While the pre-Christian Viking chieftains had supported concubines, thereby breeding many of the next generation of warriors and augmenting the number of offspring provided them by their legitimate wives, their conversion to Christianity tended to promote a stricter monogamy and to reduce the number of their offspring, without reducing the rate at which these offspring were depleted in combat. Gustavus Adolphus weakened Sweden forever in his attempt to build a Swedish Empire over much of Northern Europe, just as Napoleon depleted France in his effort to build a French-dominated empire. Ancient Greece destroyed itself by constant internecine warfare, in which the educated and intellectually distinguished aristocracy eventually eliminated themselves. Consequently the Romans had little difficulty in conquering Greece, before they too eliminated their leadership families in continued warfare.

Also, as a result of the overseas colonizing activities of the Europeans, and relative blindness of Roman Catholicism to racial differences, many individuals in Europe and North America who pass as "white" or Caucasoid have inherited genes from other populations of lower average IQ. The Northern Chinese, Koreans and Japanese have not been affected by alien genetic admixture to the extent experienced by the Caucasoid populations of Northern Africa, Western Eurasia and the Americas. This author consequently believes that the West Eurasians were once equally or more creative, and equally or more intelligent, than the East Eurasians. Early Chinese cultural progress was largely stimulated by cultural infusion (and possibly some genetic infusion) from West Eurasia via the Asian steppes; but the West Eurasians followed a more dysgenic course, particularly after their acceptance of Christianity. The influence of Confucianism and Shintoism served one way or another to keep the East

Eurasians closer to the eugenic laws of nature, and so better preserved the mental abilities with which evolutionary selection during the last Ice Age had endowed the hominids of Northern Eurasia.

We may well ask ourselves what has caused the prevailing epidemic of biological egalitarianism in the Western World – for it is an ideological disease to which other cultures seem to be more resistant. It would appear largely to originate in the ethics and philosophy of Christianity. The Judaic Old Testament repeatedly condemned those who cohabited with alien women, and so tended to protect the gene pool from admixture with "lesser peoples" not chosen by God. Christianity, by contrast, evolved as a missionary religion portraying all humankind as equal in the eyes of God. All peoples were considered to be of equal innate merit, needing only to be led to recognize the supreme authority of the Christian God. Pre-Christian Europe had believed that its best and brightest were superior beings descended from the gods; but the Christian ethic portrayed all peoples, nations and races as equal before God. The lower their level of achievement, the more deserving they were of sympathy and assistance.

Saul, when he took the name of Paul and laid the foundations of Christianity, knew that he would be recruiting his followers from amongst a heterogeneous mass of slaves and persons of low status who sought moral equality with the aristocratic *gens* that governed the Roman Empire. Beginning as a Judaic heresy influenced by the communistic ideals of the Essenes, early Christianity attracted converts by attacking the idea of pride of birth and human inequality, and portraying all peoples of all races as equally low before God.

Today the doctrine of egalitarianism dominates the Western nations with a quasi-religious mystique rooted in the notion of biological uniformity, but the resultant spirit of universal altruism is primarily restricted to the culture of the Western world. If the East Eurasians eventually achieve world dominance – and already there is evidence that this may come to be – they will surely one day reflect upon the history of the "dog-eyed" West Eurasian race, and wonder at the stupidity of a people who could uncover the secrets of biology but still destroy themselves by refusing to abandon the more dysfunctional aspects of the confused melée of dysgenic social mores they had inherited from their pre-scientific past.

X

THE HUMAN GENOME PROJECT

THE GROWING APPRECIATION of the role of genetic forces in determining human qualities has led in recent years to the extensive research into the human genome which for convenience are known simply as the Human Genome Research Project.[1] Despite the hangover of anti-biological environmentalism in the social sciences, this research is focusing the attention of the more perceptive scientists on the reality of human behavioral genetics. Breakthroughs in molecular biology have made possible the construction of a human genome map, consisting of a schematic representation of each chromosome, showing the bands and the approximate location of each of thousands of genes. Work is advancing continuously in several different countries on the construction of genetic linkage maps, with special attention to those responsible for human defects, tracing the loci assignable to specific human chromosomes. Already, a database comprising over 6,000 loci has been constructed by those engaged on the U.S. project, with work also progressing in Britain and a few other countries.[2]

---

[1] The application of genetic engineering to agricultural and other industrial uses is also encouraging the scientific world to appreciate the opportunity that is rapidly developing for humankind to plot its own destiny.

[2] This has been vastly facilitated by computers, for the development of which science owes an eternal debt to that much-maligned genius, William Shockley, who headed the small three-man team which made a breakthrough in developing the transistor and who, ironically, was castigated for urging the need to research the genetic basis of human personality in order to enable humankind to guard against dysgenic

While the human genome project already points to new strategies to diagnose, treat and prevent human diseases, and may be expected to become the basis of extensive medical benefits for posterity, its utility in revealing human ancestry should not be overlooked. It has two major implications: firstly the potential for promoting eugenic remedies by pointing to the genetic predisposition to disease as distinct from the actual occurrence of disease; and secondly, the light that it promises to throw on the nature and extent of human racial diversity. As Marc A. Lappé has said,

> The products of the genome project may throw into stark relief the paradox of a society based on the premise of equal standing at creation and one that is found to be composed of a genetically heterogeneous group of subpopulations with quantitatively different frequencies of heritable traits. As a society, we will have to ask if we can in fact collect information that reveals these individual differences and still continue to treat all individuals the same....
>
> What is clear is that even a partial picture of the genetic landscape that defines the molecular differences among human individuals will reveal more its nonuniformity than its hoped for universality.[3]

Human racial history is being clarified: as new gene sequences are revealed by the genome project it is possible to cross-relate them with detailed maps of other genetic loci in the genome. This increasingly facilitates a more complete picture of gene action and the analysis of polygenic traits, which are so important to any true picture of the character of distinctive human breeding populations. As the history of

---

decay.

[3] "Justice and the Limitations of Genetic Knowledge," Marc A. Lappé, in *Justice and the Human Genome Project*, editors T.F. Murphy and Marc A. Lappé, California University Press, Berkeley, 1994.

genes is traced, the genetic profile of different breeding populations is becoming sharper.

The human genome project helps to remind us that we are, after all, biological organisms just like other animals, and even plants. It should make us realize that we need to bring ethics into line with biological reality – into line with the reality of evolution and the universe. It is essential that human culture and human ethics should be in harmony with the realities of the universe, and those realities include our biological nature. It should also serve to remind us that racial or genetic differences are real and important, and that the culture on which we rely for survival is itself dependent on our biological and genetic constitution.

The uniqueness of each individual's genetic makeup is thus today well-established. DNA fingerprints are so distinctive that the likelihood that an individual member of even that widely heterogeneous variety of populations contained within what we call the Caucasoid race may have the same DNA imprint as another is something in the region of one in a hundred million, and between individuals of different races this possibility becomes virtually zero. However, to date less publicity has been given to the significance of group differences in the frequency of major genetic loci.

Contrary to previous egalitarian supposition and propagandizing, significant genetic differences distinguish historically isolated populations or gene pools. A prime question for the future is the extent to which historically separate populations, some of which have evolved further in certain directions than others, will be able to preserve their distinct genetic attributes in an increasingly overcrowded and mobile world. Evolution was able to proceed amongst hominids, as amongst other life forms, as a result of genetic isolation – which was initially facilitated by geographic isolation. Only after geographic isolation had created a wide variety of subspecies and quasi-subspecies possessing markedly different genotypes did subspecific or "race" prejudice – what

G. G. Simpson (1971, 1980) perceived as a psychological restraint on cross-breeding – come into being as an evolutionary mechanism serving to preserve subspecific identity amongst increasingly mobile species.[4] The co-habitation of individuals of diverse genetic character in the same geographical location, and the probable further decline of "race prejudice," may reverse the prehistoric evolutionary trend toward subspeciation and initiate a reversal of nature's evolutionary tactics.[5]

Opposition to the Human Genome Project

Biological egalitarians are fearful that the knowledge acquired as a result of the human genome project will destroy the plausibility of their universalist ideology. They are particularly concerned about the revival of scientific interest in group differences which will unavoidably result from the construction of 'demic maps' tracing the genetics of regional populations of different evolutionary ancestry – since these will reveal the true extent of group differences in genetic potential. Such maps will trace the varied genetic heritage of human demes (breeding populations or micro-races) with scientific precision; and by throwing a sharp and revealing light on what had hitherto been a debatable area of

---

[4] Simpson points out that without some mechanism to discourage interbreeding (what we call "race prejudice") no subspeciation could have occurred among more advanced species whose mobility rendered geographical barriers inadequate to sustain evolutionary speciation. Since speciation is one of the major factors in evolutionary advancement, some kind of psychological barriers had to develop to prevent new experimental species from being eliminated at the subspecific stage by re-merging their specific collation of genes with that of other subspecies. In the absence of biological or culturally inspired prejudices, "miscegenation" would eliminate new subspecies before they had had time to become so genetically differentiated from each other that they had lost the biological ability to interbreed.

[5] See also J. Philippe Rushton 1989, cited earlier, for an update on this argument.

human knowledge will explode the anti-evolutionary myth of human biological uniformity.

Many egalitarians apparently regard equality before the law as an inadequate basis for societal justice, and are prepared to condemn limitless numbers of future generations to the evils of ignorance in order to protect the egos of those who are alive today. Egalitarian activists such as Barbara Faye Waxman have *complained* that the human genome project will foster a revival of negative eugenics. It was not altogether surprising, therefore, that, when in 1988 the European Commission proposed the establishment of a human genome project to aid predictive medicine, an anti-eugenic coalition immediately sprang into existence and succeeded in inducing the European Parliament to establish a committee to report on the proposal.

The rapporteur appointed for the committee was a certain Benedikt Härlin – a member of the Green Party, which has strongly opposed genetic research and the genome project. The Härlin report consequently recommended that tight restrictions be imposed on scientific research into human genetics. The Green Party, for all its laudable claims to protect the non-human environment, turns a blind eye upon the destruction of the human genetic heritage, arguing that the prime responsibility of the European Parliament is to block any eugenic trends that might result from human genome research. The ultimate result was that the project was severely emasculated, narrowing the opportunities of securing the information so necessary for predictive medicine, prohibiting both human germ line research and eugenic intervention – with a portion of the approved money being diverted to fund "ethical issues arising from genome research." Cynics might well see this latter as a carrot offered to what Bernard D. Davis called the "scientific Luddites".[6]

---

[6] See Davis's *Storm Over Biology* already cited, and also chapter 11, footnote 8.

XI
EUGENICS AND MODERN MEDICAL SCIENCE

IT IS NOW SIXTY YEARS since the Nobel-prizewinning geneticist Hermann J. Muller confirmed the view of the earlier eugenicists that humanity is threatened by massive dysgenic trends due to the relaxation of natural selection. To quote Muller's own words, modern humankind is being biologically undermined:

> For civilization affords an excellent chance to breed for many sorts of the weak, the stupid, and the vicious characters which are continually arising ... So long as this condition holds, then, a biological disorganization will take place, and will necessarily continue without limit, or until the species disappears ... under civilization many traits that are detrimental to the species as a whole are propagated at a *faster* rate than the normal ...[1]

THE IMPACT OF MODERN MEDICINE

TODAY, PROGRESS IN MEDICAL SCIENCE is facilitating effective and humane eugenic practices, while at the same time promoting dysgenic trends.

On the positive side, genetic analysis is transforming medical science by facilitating new opportunities for the remedial treatment of genetic defects and by offering the possibility of reducing human suffering by identifying fetuses afflicted by severe, incurable defects at the prenatal stage. The genetic roots of expanding numbers of illnesses are being identified, and family medical history records have become

[1] Hermann J. Muller, *Out of the Night* 1935, p. 42-43.

increasingly helpful in assisting medical researchers to determine appropriate courses of treatment. Such records are not always available, however, and the knowledge that can be gained therefrom is far from definitive as an indicator of genetic weaknesses in the individual.[2] Eugenic decisions are more precise when the phenotype of the infant can be examined by means of techniques such as ultrasound sonar examination. By this process an ultrasound transducer is placed on the mother's abdomen and sound waves travelling through the amniotic sac will reveal the size and shape of the fetus within the womb. In addition, laparoscopy has been developed, which involves the use of a flexible fiberscopic rod, inserted through the abdomen, which in addition to permitting an examination of the abdominal cavity and fallopian tubes, makes it possible to obtain blood samples from the fetus which can then be analyzed for evidence of genetically determined diseases.

Another technique utilized today in promoting practical eugenic treatment is amniocentesis. This involves the extraction of a small quantity of amniotic fluid, which carries in it cells shed by the fetus, from which, in turn, the genetic make-up of the fetus can be determined. Already, some genetic defects can be cured, while others may be deemed sufficiently serious to indicate the desirability of terminating the pregnancy.[3]

---

[2] The Mormons are exceptional in the excellence of their genealogical records. These have proved invaluable research aids to gene hunters. Some people have elected to undergo preventative mastectomy solely on the basis of family medical history, but Myriad Genetics of Salt Lake City is currently developing tests to identify the genetic factors responsible for breast cancer which if successful will render such scatter shot measures unnecessary.

[3] *The Washington Post,* September 17, 1995 (p. A3), carries a story entitled "After Years of Birth Defects, Pain and Mystery Linger in Brownsville." Author Sue Anne Pressley reports on the high rate of neurological birth defects in Brownsville, Texas, caused by air pollution from factories in the neighboring Mexican town of Matamoros. The

However, those concerned about the need to transmit a healthy gene pool to future generations do not forget that there are two sides to advances in medicine: improved medical skills may help the individual to overcome genetic handicaps and are to be commended on compassionate grounds, but at the same time they often heighten the chances that the genetic problems responsible for the condition will be passed on to future generations. Medical science does not benefit humanity when it has a dysgenic impact – when it defeats nature's normal tendency to reduce the likelihood of harmful genes being transmitted in perpetuity to future generations.

Sir Julian Huxley warned against this in 1951:

> we can be certain on theoretical grounds that the relaxation of selection brought about through our medical knowledge and social care must be causing a slow degeneration of the stock, through the accumulation of harmful mutations.[4]

Today the "genetic load"[5] of harmful genes is widely

---

article reports that among the many serious birth defects recorded in Brownsville between 1986 and 1991, 47 involved babies born without brains. The ability to detect this condition in fetuses now facilitates eugenic abortions, which spare the parents and the community much emotional and economic hardship. While the genetic defects responsible for anencephalic births are hardly likely to be passed on to future generations, many other defects of a less crippling but almost equally disastrous effect often do enter the gene pool of a population and increase the genetic load of hereditary defects which will be passed on to posterity.

[4] J. Huxley in his Introduction to *Human Fertility* by Robert C. Cook, 1951. New York: Sloane.

[5] The term "genetic load" was popularized by Hermann J. Muller, who repeatedly warned that: "[I]n a state of nature, this problem of the accumulation of defective genes was often taken care of by 'natural selection' ... The only adequate alternative to it, however, is some kind of thoroughgoing eugenics ... Failing this, a biological deterioration of the

spread throughout the outbreeding populations of the modern Western world, and many of those who inherit the finest collection of genes must necessarily be cursed by the inclusion of one or more dysfunctional genes. As Huxley warned, modern medicine increases the likelihood that harmful genes will be transmitted to future generations; but the permeation of our society with such genes is now so advanced that the earlier recommendation of eugenicists that even intelligent and otherwise healthy individuals who might be carrying a solitary deleterious gene should refrain from reproducing is no longer appropriate today. In consequence it is imperative that geneticists should be encouraged – not discouraged – to search for techniques which will make it possible to substitute healthy genes for faulty ones. Genetic surgery and test-tube babies are nothing to be frightened of if they are the price that must be paid to preserve posterity from genetic pollution. Unless blind emotions are permitted to prevent scientific investigation in this direction, techniques for genetic surgery which will make it possible to eliminate various genetically-transmitted conditions will surely be developed in the foreseeable future, and research in this direction is of imperative urgency.

The Enhanced Potential for Positive Eugenics

Both eugenic approaches, "negative" (the reduction of harmful genes in the gene pool) and "positive" (an increase in the percentage of "healthy" genes in the gene pool), are currently being greatly facilitated by modern science, and an increasing number of medical personnel are already making eugenic choices in the conduct of their duties. When eugenics

---

human race seems inevitable, and this might even endanger the dominance of man as a species." (*Out of the Night*, p. 44). In his article "Our Load of Mutations," published in the *American Journal of Genetics*, June 1950 2:2, p. 165, he wrote, "In the case of mental traits even more than those having to do with general health and vigor, there is reason to conclude that selection has greatly relaxed under modern conditions."

first became popular, genetic knowledge was too limited to facilitate negative eugenics in any but the most crude manner, and much was done that was tragic and repulsive. Without debating the history of legally enforced sterilization, there were also many instances where individuals of superior genetic quality were persuaded not to have offspring where there was reason to believe that a solitary defective gene had revealed itself in the phenotype of a family member. Today there is an increasing ability to detect genetic defects in fetuses, and the future holds the possibility of "genetic surgery" to replace deleterious genes with healthy genes, thus making it possible to free future generations from certain specific weaknesses in a discriminating fashion. Competent individuals whose genotype is cursed by only simple genetic problems may in future be able to contribute their superior genes to posterity without fear of transmitting genetic defects which are identifiable and subject to manipulation by such techniques. We can now look forward to the days when the breeding of reliably more healthy and competent stock can be promoted by ethically sound medical intervention.

Continuing progress in genetic science holds out the promise that gene 'splicing' or genetic 'surgery' will make the elimination of many hereditary defects a real possibility in the foreseeable future. It would then be possible to replace individual deleterious genes by healthy counterparts so that future generations can inherit all the desirable qualities of their forebears free from adverse mutations and other heritable disabling conditions. Yet even this will not solve the total dysgenic threat. To take an analogy from mechanical engineering, it is easy to repair a disabled Rolls Royce by replacing a leaking hosepipe or other defective part, but it is not practicable to attempt to convert a Yugo or Lada into a Rolls Royce by replacing a major number of components. The task is far too large, and many of the Rolls Royce parts would be incompatible with those of the inferior vehicle. The justification for "positive" or "mainline" eugenics – the encouragement of an appropriate rate of reproduction among overall healthy and competent individuals – would therefore still remain, even after effective systems of genetic splicing had been developed. The old ideal of the mainline eugenicists – that of ensuring an adequate reproduction of the "Rolls Royce" models of humankind – remains a pressing necessity. As one cannot emphasize too often, each generation is a genetic bottleneck, and the failure of any generation to replicate its better biological specimens means that their genes are lost forever and for all time.

As another example of what can be done in this respect, *in vitro* fertilization has already become common practice in Western countries when married couples who wish to have children are prevented from doing so by the infertility of one of the prospective parents. Such situations offer a perfectly moral opportunity to influence the course of human evolution in a rational and eugenic manner, by ensuring that the couple should have available to them a choice of germinal donors of healthy and genetically sound stock. It was Hermann J. Muller[6] who first widely publicized the idea of establishing repositories for germinal plasma as a means of maintaining eugenic quality; and it was Dr. Robert K. Graham, the distinguished optometrist who contributed to the welfare of humankind by developing the scratch-resistant plastic lens – and also wrote an inspiring pro-eugenics text entitled *The Future of Man* (1970) – who converted Muller's idea into a reality. Aided by an advisory panel of medical specialists and geneticists, Graham established the Repository for Germinal Choice in Escondido, California, in 1980. Here, would-be parents who are unable to conceive children of their own due to the infertility of the husband can obtain germinal plasma of high quality after perusing information about the physical and mental qualities of the donor. Up to the time of writing, the Repository has been responsible for the birth of over two

[6] See Hermann J. Muller *Human Genetic Betterment*, 1963, New York: American Humanist Association.

hundred bright, lively, and healthy infants.[7]

CRITIQUE OF THE EGALITARIAN OPPOSITION TO EUGENICS

THE MORALITY of even this modest eugenic technique does not prevent the "scientific Luddites"[8] from continuing to condemn even the mildest of voluntary eugenic programs such as this, seeking to ridicule such programs at every opportunity. Many of the opponents of genetic research today accept the logic of eugenics, but do not care what happens to future generations. They seemingly prefer to overlook the words of the young woman who, after discovering from genetic screening that she was free of the gene for Huntington's disease, said that:

> After 28 years of not knowing, it's like being released from prison. To have a hope for the future ... to be able to see my grandchildren.[9]

Unfortunately, while knowledge can help us to perceive the human condition more clearly and to evaluate the results of our behavior, people do not always act rationally. As J. Murphy observed (1982, p. 100), the ends that individuals accept as ultimate goals "are given by the passions and, at this level, reason is the slave of the passions." Only a passion that

---

[7] An account of the Repository has been provided by Jamieson in an article published in *The Mankind Quarterly* under the title "Concerning Scholarly Creativity: Hermann J. Muller and Germinal Repositories" (Jamieson, 1993).

[8] Bernard Davis (1986) likened the opponents of genetic research to the Luddite rioters of eighteenth century England, who tried to halt the industrial revolution by breaking newly installed machinery. Like them, scientific Luddites oppose the search for knowledge even though it could free future generations from the dysgenic impact of the faulty social policies of today. They fail to see that policies rooted in ignorance are far more dangerous for the future of mankind than policies based on factual knowledge.

[9] Jerry E. Bishop and Michael Waldholz, 1990, p. 274.

takes into consideration the welfare of those yet to be born is justifiable in moral terms. It is the welfare of the entire phylogenetic continuum, not that of the individual alone, that should shape our passions and our morality.

Apart from the obvious outward physical differences which distinguish the diverse human breeding populations, behavioral tendencies which have major importance for determining the viability of civilization and for the better scientific understanding of the universe also seem to have a predominantly genetic basis. We do not here refer solely to intelligence, of enormous significance as all must surely concede that quality to be. The ability to direct one's actions toward achieving long-term goals, rather than to yield to the siren call of instant gratification, is another of the many qualities that are of supreme importance to modern man. General physical health and the propensity for properly directed altruistic behavior are other desirable qualities which also have strong genetic roots. On the other hand, specific diseases and sociopathic behavior often have a genetic basis, and call for the attention of eugenicists. The genetic basis of such behavioral traits is likely to be complex or polygenic, but the study of animal models as well as of human twins has shown that all behavioral characteristics have a genetic basis, and that the environment can only work on the material provided by heredity, occasionally modified by mutation.

The argument against eugenics is based on the clearly erroneous assumption that genes are spread evenly throughout human populations, or in more extreme cases, on the idea that no one gene is more desirable than another. Such views would imply a degree of genetic panmixia which could not have existed in evolutionary time; otherwise evolution could not have occurred. It has been said that when a species is reduced to a single subspecies (e.g. panmixia), it is nearing extinction. Long-term evolutionary survival is by way of speciation and this necessarily involves subspeciation. Evolution cannot occur unless "favorable" genes are segregated out from amongst "unfavorable" genetic formulae.

The entire history of evolution from biological simplicity to biological complexity and from biological uniformity to biological diversity involves genetic segregation just as it involves genetic selection. Furthermore, human behavior is shaped by environmental forces operating on the genetically determined potential, and we should remember that even *the capacity to learn* from accumulated culture is genetically influenced.[10]

Examples of legislation that has been successfully introduced to suppress the utilization of genetic science can be found in many states. Genetic screening as a basis of employment is illegal in some states, such as Wisconsin and Iowa [Iowa Code Ann. 729-6 (West 1992), Wisconsin Statutes 111.372 (1991-1992)], while a Louisiana Statute (La. Rev. Stat. #9: 122) rules that "a viable *in vitro* fertilized human ovum is a juridical person which shall not be intentionally destroyed."[11]

It is entirely true that the opportunity to interfere with nature – in the willful selection by parents, for example, of the sex of their offspring – could lead to serious social problems if foolishly applied; but, on the whole, knowledge must be deemed always to be desirable, and the search for knowledge about human heredity should not be suppressed or discouraged at the whim of the scientific Luddites. Many societies with advanced cultures have long tended to allow unnatural cultural selection to supplant natural selection, often with seriously dysgenic results, and the question of how humankind will choose to use the expanding corpus of genetic knowledge is a sociological and philosophical problem – not an argument for suppressing knowledge. Science has entered upon a new phase in the study of genetics, one which has wholly positive implications for the future of mankind –

---

[10] See Beckstrom, 1993, p. 14.

[11] Interestingly, the legislators did not prohibit the freezing of such embryos, which in effect renders them non-viable.

provided dysfunctional ethical and cultural forces do not prevent the utilization of this knowledge and thereby deprive future generations of their right to a healthy and creative existence.

XII
GENETICS AND THE FUTURE OF HUMANKIND

THE RAPID PROGRESS in medical and genetic research, not to speak of the human genome project, is now exposing the fallaciousness of the assumptions on which biological egalitarianism was raised just a hundred or so years ago. Indeed, future civilizations will look with amazement on the absurd notions that have distorted the minds of Western men and women in the Twentieth Century. The idea that the individuals that comprise any society – particularly a multi-racial society, which by definition will be genetically more heterogeneous than one comprised of individuals who share a common ancestral heritage – could be of uniform biological potential is exposed as a fabrication and a logical absurdity. Equality before the law remains a realistic social goal, but equality of ability can never be a biological reality, except between identical twins. Some individuals – especially in a more heterogeneous society – will possess genetically determined mental and physical qualities which will enable them to be more creative, more productive, or physically more healthy than others, even when the cultural or environmental playing field is level. If that were not the case, there could have been no possibility of evolution.

Steven Jones, a geneticist at University College, London, has in recent years warned against the possible reversal of human evolution in his book *The Language of Genes: Solving the Mysteries of our Genetic Past, Present and Future* (1994). He comments that "Natural selection has to some extent been repealed" with modern societies seeming "to conspire to slow down human evolution." Under natural selection the fit tended to survive to produce offspring while the unfit failed to leave the same proportion of progeny. But modern social conditions have reversed that tendency. He was in effect repeating the all- too-realistic warning of the socialist George Bernard Shaw, who ninety years earlier had warned that

> ... there is now no reasonable excuse for refusing to face the fact that nothing but ... eugenics ... can save our civilization from the fate that has overtaken all previous civilizations.[1]

William K. Stevens recently authored an article aptly entitled "Evolution of Humans May at Last be Faltering".[2] Rapid evolution occurred during the Pleistocene, especially in those more temperate areas affected by the Ice Ages, and this was facilitated by the isolation of relatively small groups in disparate geographical environments. However, the barriers which formerly enabled specific populations to evolve rapidly under severe selective pressures, and then spread outward to replace less advanced subspecies, have now been removed by modern technology and the vastly increased size of the overall hominid population. As a result a melding of populations is creating a global gene pool, which by its very size reduces the chances of further evolutionary selection – quite apart from the averaging down of intelligence as the more intelligent either interbreed with the less intelligent or are outbred by them. The belief that humankind could benefit from being levelled into a single subspecies flouts the basic principle of evolutionary advancement, which is rooted in differentiation.

Concerned about this aspect of the problem of world overpopulation, Timothy H. Goldsmith, the Andrew W. Mellon Professor of Biology at Yale University, maintains that "evolution enters science with biology, but it is an exercise in

---

[1] George Bernard Shaw, *Sociological Papers,* pp. 74-75. London: Macmillan & Co Ltd. 1904.

[2] *New York Times*, March, 14, 1994

self-deception to suppose that it leaves the cast when the social sciences come onto the stage."[3] Culture can distort value systems, he warns, and

> [T]he concept of morality's producing the greatest good for the greatest number is consistent only when the interests of the individuals are very similar. This has probably frequently been the case with small homogeneous groups in competition with other groups; it is less obviously so when groups are large and heterogeneous.[4]

William K. Stevens further quotes Ian Tattersall, a paleoanthropologist at the American Museum of Natural History, who warns that "*Homo sapiens* today is in a mode of intermixing rather than of differentiation, and the conditions for significant evolutionary change simply don't exist – and won't, short of some all-too-imaginable calamity."[5]

"KNOW THYSELF"

G. G. Simpson early condemned the popular assumption that evolution necessarily involves a constant enhancement of morphogenetic or phylogenetic "fitness to survive." Evolution as he defined it is simply morphogenetic change – which can sometimes be dysgenic and lead to extinction. The long term survival of any one phylogenetic continuum is by no means

---

[3] *The Biological Roots of Human Nature: forging links between evolution and behavior*, Timothy H. Goldsmith, p. 141. Oxford University Press, New York, 1991. See also *Evolution and Human Kinship*, 1988, Oxford University Press.

[4] Ibid. p. 123.

[5] Yet science offers a multitude of solutions to the world's problems. In this age of unprecedented population growth, and the invasion of national breeding grounds by the surplus population of alien stocks, illegal immigration could be readily detected through the use of identity cards impregnated with small portions of the carrier's own unique DNA. This would establish a forgery proof method of identification, as well as provide a means of classifying the phylogenetic identity of the individual.

guaranteed. As he warns, "to a geneticist fitness has nothing to do with health, strength, good looks or anything but effectiveness in breeding [in the short-term prevailing circumstances]."[6] Writing of the contemporary Third World population explosion, which was already apparent thirty years ago but has reached ecologically-threatening proportions since then, he observed that:

> (Latin Americans ...) Asians and Africans are likely to increase proportionately while white Europeans decrease. These evolutionary changes are bound to be accompanied by political changes, changes in institutions and forms of government, and acute stresses on the whole social fabric.
>
> ...Either this population increase will be stopped by some actions of man himself, or it will end by man being packed elbow to elbow ... If there are no strong barriers to interbreeding in such a large population, evolution [genetic change] can be expected to be extremely slow ... as to the directions of such [genetic change] ... it could hardly be the more intelligent who would have high birthrates ...[7]

"Multicultural" societies seldom result in the communal bliss envisaged by the proponents of multiracial integration and biological egalitarianism. Stresses follow naturally, despite the most ardently concerted efforts to promote harmonious race relations.[8] At the one extreme we may consider Brazil, long used as an example of a country where racial admixture between the various races is far advanced: Brazil demonstrates the inevitable segregation of the more intelligent from the less intelligent, with resultant extremes of wealth and poverty.

---

[6] C. G. Simpson, *This View of Life: the world of an evolutionist*. 1964, p. 273. New York: Harcourt, Brace and Jovanovich

[7] ibid, p. 282-283

[8] See Lazlo F. Thomay, *The Natural Law of Race Relations, 1993* (Scott-Townsend Publishers, Washington D.C.) on this issue.

Peoples of largely European ancestry (joined more recently by Asian immigrants) predominate at the upper end of this vastly extended spectrum of wealth and influence, while those with a higher percentage of African and Amerindian genes predominate at the lower end.

The United States of America today stands as an example of a multi-racial society where racial admixture is as yet less advanced than in Brazil. Yet in the U.S.A. it is only vast government-legislated transfers of wealth from the more successful ethnic groups to the less successful minority elements that prevent the majority of the latter from sinking into the same dire poverty which afflicts the Brazilian masses. Essentially, the problem is that as technology advances there are fewer work opportunities for those who lack the intellectual ability to handle tasks which are not primarily physical. Fewer and fewer work opportunities are available for those whose intellectual powers are low but quite adequate for laboring tasks in horticultural or pre-industrial agricultural societies. An examination of the social and economic problems facing the United States in its increasingly multiracial condition may be found in Jared Taylor's *Paved with Good Intentions* (1992).

Emotionally-inspired but misguided and ineffective responses to the problem of world over-population, with its concomitant pressures promoting human migration and environmental destruction can only be harmful to posterity – not beneficial. Humankind needs to heed the ancient Greek injunction to "Know Thyself." An all-embracing pan-species form of altruism – which Huxley would regard as a perversion of the natural role of altruism – might still be workable from an evolutionary point of view if it were motivated by *truly* altruistic emotions which extended beyond the present generation to embrace as its goal the well-being of all those untold generations yet to be born. Yet to be born? Which *might* be born provided humankind does not wreck the biosphere or destroy the genetic heritage created by some four billion years of natural selection. We must direct our concern not just to the well-being of those who live at this present time, but to the well-being of that far greater number who are yet to follow after the present generation has passed away. We can do no greater good than to leave the multitude of future generations a healthy genetic inheritance in a healthy and unspoiled environment. If we care for the future of humankind then we must breed preferentially from those most likely to bear offspring capable of lifting humankind to greater heights, not from those who would drag it downwards, back into the murky past. As our knowledge about the universe in which we live expands, it becomes increasingly obvious that the survival of humankind will depend on improving the genetic endowment, not averaging it steadily downward.

Clearly, rational altruism requires each generation to avoid pollution of the ecosphere and of the human genetic heritage. Referring to the misguided sense of altruism which governs contemporary Western civilization, Lord Justice Lawton of Britain remarked that a new Edward Gibbon, writing about the history of the West, might well entitle the last chapter "The Age of Compassionate Fools". As a people, he observed, we have allowed a misguided idealism to override common sense.

Most of those who have sought to suppress human knowledge about heredity have done so with kindly intentions, but sound policies can never be constructed on bad science or unsound data. Any society that sets itself against the immutable causal laws of biology and evolution will ultimately bring about its own demise. An inappropriate culture that is not in harmony with the mechanics of the universe has only a poor chance of surviving. Heavily dysgenic trends have dominated this century as a result of the selective elimination of talented personnel – such as the heavy loss of life among air crews involved in modern warfare, the genocidal slaughter of the elite in the Soviet Union and Maoist China, and the general tendency for the more creative members of modernized societies around the world to have fewer children than the less creative. Nor should one ignore the impact on mankind of the increasing destruction of the biosphere by the wanton use of harmful technologies and by the unprecedented population explosion which is devastating the natural environment in even the poorest and least technologically advanced countries. Nevertheless, if the world does not collapse beneath the triple pressures of pollution, overpopulation, and severe dysgenic trends, and if governmental policies could be modified to take into consideration the findings of a scientific community encouraged to conduct human behavioral genetics research uncensored by radical Luddite ideologues, it is possible that the human race might still have a future worth working for.

But Will Eugenic Policies be Accepted?

The need for some form of eugenic intervention, humane and benevolent, is today increasingly urgent due to the fact that human culture and human technology have blunted nature's biological pruning knife. At all times a degree of genetic selection is necessary if any breeding population, subspecies or even species is to survive, since mutations (mostly harmful) occur continuously. Medical intervention is unquestioningly humane when it assists those born with genetic handicaps to cope with them, or those who suffer from accidental handicaps to survive; but eugenic goals make it highly desirable that research aimed at advanced techniques of "genetic surgery" be developed to enable the substitution of healthy genes for deleterious genes. Too many individuals in contemporary society, of high genetic potential, now carry one or two undesirable genes for us to believe, as the early eugenicists did, that anyone with a known genetic defect should simply not reproduce. Yet research aimed at developing techniques for replacing faulty genes with healthy ones is the one area of study which the opponents of eugenics oppose the most stringently.

To foretell the future of eugenics, the question we must ask is: how will human societies react once the full truth about individual and group genetic ability becomes known? Unfortunately, nowhere is it written that a society's ethical views or any other aspect of its cultural baggage should have eugenic value. Only in the long term can we say that dysfunctional ethico-cultural systems are doomed to extinction, simply because if they persist they will doom the biological communities which are guided by them. There is only one law in the universe: that which does not survive ceases to exist. Any ethic which has overall dysgenic implications will eventually eliminate the society which adopts it. While altruism evolved as an evolutionary device that promoted the survival of the specific gene pool which gave rise to it, any population that adopts a perverted or dysgenic form of altruism – one which encourages a breeding community to breed disproportionately from those of its members who are genetically handicapped rather than from those who are genetically favored, or which helps rival breeding populations to expand while reducing its own birth rate – is unlikely to have much of a future.

Once science can determine which individuals and populations have disproportionately high gene frequencies for advantageous traits, and which are handicappedby deleterious genetic qualities, societies which are prompted by a higher altruism which embraces posterity will be able to effectively select eugenic over dysgenic reproduction. If true altruism prevails, the result will be eugenic decisions, made voluntarily and without coercion. By contrast, if the ethical level is low (a categorization which includes dysfunctional ethical systems, however ardently such unrealistic value systems might be held by individuals who regard themselves as moral), or the altruistic drive is diverted into unnatural directions, then the choices made are likely to be dysgenic.

Forced, state-imposed eugenic systems are wisely feared by many well-intentioned people for good reason: human governmental bureaucracies seldom achieve the virtue anticipated by high-minded philosophers such as Plato, and tend too often to foster corruption and to permit unworthy individuals to seize positions of power and influence. Indeed, the leadership of most if not all nations, intellectual as well as political, appears to have currently fallen into the hands of individuals and pressure groups which promote dysgenic policies that match, in their impact upon humanity, the overall destruction of the ecosphere which is currently taking place around the world, and which few governments or interest groups seem capable of preventing. Possibly the best that we can hope for is that an intelligent and informed populace, freed from the distorted myth of biological egalitarianism and any belief in the near-divinity of humankind as the "apple of God's eye", may exert a realistic and humane influence on public policy concerning the critical problems today facing humanity.

THE ALTERNATIVE: DISASTER

J. STAN ROWE, in a final chapter to *Planet Under Stress: The Challenge of Global Change,* writes that if people are regarded as being the central reality of the universe

> ... then the sole focus of ethical concerns will be *their* rights and values ... but if things other than humans are of surpassing importance, as today's events lead many of us to suspect, then a new purpose – Salvation of this Planet – is revealed ... [science] has inadvertently opened a new chapter in the book of knowledge whose dazzling insights can change fundamental ideas about the planet/people relationship, about values, and about the way we know the world.[9]

The present author shares the same concerns as Rowe, but would stress that logic should make it perfectly clear even to the most anthropocentric of philosophers that the fate of humankind itself requires that we recognize the evolutionary

---

[9] J. Stan Rowe, *Planet Under Stress: The Challenge of Global Change*, Editors, Constance Mungall and Digby J. McLaren, p.332.

realities that govern the workings of the universe – a reality in which humankind plays only a very subordinate and minor role, and in which the laws governing evolution reign supreme.

This author summarized his views on the vital issues now at stake, resulting from the reversal of natural selection – due in no small part to dysgenic social ideologies – in the concluding chapter of his book *Race, Intelligence and Bias in Academe* (1991):

> If truth prevails in our society, social science and social and political philosophy will eventually be obliged to take cognizance of the fact that human behavior is subject to the laws of causality, as is all matter throughout the universe. It is now clear that genetic factors set the potential limits of behavior by the human animal, while environmental circumstances and events influence the individual organism within the potential limits of its behavior as determined by heredity. Not that one can say that either is more important than the other. Environment and heredity are two different categories of causal factors which intertwine and interact. Even to ask which is the more important can be a misleading question - just as it is to ask which is more important, the cart or the horse?
>
> Yet there is a vitally significant difference between genetic forces and environment, a difference which raises eugenic considerations to a level of superior concern. One can enhance the environmental component of intelligence by creating a more favorable environment for the next generation, but this is only a temporary boost, and must be maintained at the heightened level for each successive generation. Its effect is not intergenerational as the Lysenkoists believed. By contrast, an improvement in the genetic constituency of a population is permanent, unless eliminated by new dysgenic forces. A genetic improvement is an intergenerational improvement, and genetic or dysgenic decline represents a genetic loss

> which is likewise transmitted intergenerationally. Unfortunately, a single generation of severe dysgenic decline can destroy tens of thousands of years of evolutionary progress laboriously achieved by means of natural selection – and thus constitute a permanent blight on posterity.

Every generation is a genetic bottleneck, holding in its care the genetic and environmental heritage bequeathed to it by past generations. Any species that adopts patterns of behavior that run counter to the forces that govern the universe is doomed to decline until it either undergoes a painful and totally involuntary eugenic process of evolutionary reselection at the hands of Nature, or else is subjected to an even more severe penalty – extinction.

REFERENCES

Andrews, Lori B.
1987 Medical Genetics: a legal frontier. Chicago: American Bar Foundation
1992 "Torts and the Double Helix: malpractice liability for the failure to warn of genetic risks." *University of Houston Law Review*. 29:149-184

Baker, John R.
1974 Race. London: Oxford University Press

Beckstrom, John H.
1993 Darwinism Applied: evolutionary paths to social goals. Westport, CT: Praeger

Bell, Alexander Graham
1914 "How to Improve the Race." *Journal of Heredity* 5 (1)

Benedict, Ruth
1934 Patterns of Culture. Boston: Houghton Mifflin
1940 Race: Science and Politics. New York: Modern Age Books

Benedict, Ruth, and Gene Weltfish
1943 The Races of Mankind. Public Affairs Pamphlet no. 85. New York: Public Affairs Committee.

Binder, Rudolph M.
1915 "Eugenics and Religion". *Eugenical News* 18 (2)

Bishop, Jerry E., and Michael Waldholz
1990 Genome: The Story of the Most Astonishing Adventure of our Time – The Attempt to Map All the Genes in the Human Body. New York: Simon and Schuster. p. 274

Boas, Franz
1911 The Mind of Primitive Man. New York: Macmillan.
1928 Anthropology and Modern Life. New York: Norton.
1940 Race, Language and Culture. New York: Macmillan

Bogardus, Emory S.
1928 Immigration and Race Attitudes. New York: David McKay
1960 The Development of Social Thought. 4th edition. New York: David McKay
1967 A Forty-year Racial Distance Study. New York: Cooperative League of the U.S.A.

Bouchard, Thomas J. Jr., and Matthew McGue
1990 "Genetic and Rearing Environmental Influences on Adult

Personality: an analysis of adopted twins reared apart". *Journal of Personality* 58:1. March

Boyd, William C.
1950 Genetics and the Races of Man: an introduction to modern physical anthropology. Boston: Little Brown.

Brigham, Carl C.
1923 A Study of American Intelligence. Princeton: Princeton U. Press

Caton, Hiram
1990 The Samoa Reader: anthropologists take stock. Lanham, MD: The University Press of America

Cattell, R. B.
1937a The Fight for our National Intelligence. London: King
1937b "Some Relations of Intelligence to Socioeconomic Factors" *Eugenics Review.*
1972 A New Morality from Science: Beyondism. New York: Pergamon Press
1982 The Inheritance of Personality and Ability. New York: Academic Press

Cavalli-Sforza, Luca C., P. Menozzi and J. Mountain
1988 Reconstruction of Human Evolutionary Biology: bringing together genetic, archaeological and linguistic data. *Proc. Natl. Acad. Sci.* 85: 6002-6006

Cavalli-Sforza, L. Luca, Paolo Menozzi and Alberto Piazza
1994 The History and Geography of Human Genes. Princeton, NJ: Princeton University Press

Clark, Kenneth B.
1947 "Racial Identification and Preference in Negro Children," *Readings in Social Psychology*, pp. 174-5. Newcomb and Harley, Eds., 1947

Comas, Juan
1961 "Racial Myths". In *Race and Science.* New York: Columbia University Press

Comte, Auguste
1896 Positive Philosophy. Translated by Harriet Martineau. 3 vols. London: George Bell and Sons

Coon, Carleton S.
1954 The Races of Europe. New York: The Macmillan Company
1962 The Origin of Races. New York: Alfred A. Knopf
1965 The Living Races of Man: Alfred A. Knopf

Darlington, C. D.
1958 "The Control of Evolution in Man". *The Eugenics Review* 50:3
1964 Genetics and Man. London: George Allen and Unwin

1969 The Evolution of Man and Society. London: Lowe & Brydone

Darwin, Charles

1859 The Origin of Species by Means of Natural Selection or the Preservation of Favored Races in the Struggle for Life. Reprinted 1914. Philadelphia: Richard West

1871 The Descent of Man. London: John Murray

Darwin, Sir Charles Galton

1951 The Next Million Years. London: Hart Davis

1960 "Can Man Control his Numbers?" *The Mankind Quarterly* 1: 2

Davenport, Charles B.

1919 Heredity in Relation to Eugenics. New York: Henry Holt

Davis, Bernard D.

1986 Storm over Biology: essays on science, sentiment and public policy. Buffalo: Prometheus Books.

Dorit, R. L., H. Akashi and W. Gilbert

1995 "Absence of Polymorphism at the *ZFY* Locus on the Human Y Chromsome". *Science* 268:1183-1185. May 26

Ehrlich, P. and S. Felman

1977 The Race Bomb. New York: Quadrangle

Ellis, H.

1912 The Task of Social Hygiene. New York: Houghton Mifflin

Emerson, Ralph Waldo

1929 The Complete Writings of Ralph Waldo Emerson. New York: Wm. Wise & Co.

Eysenck, Hans

1971 Race, Intelligence and Education. London: Temple Smith

1975 The Inequality of Man. Edits Publications

Fairchild, H. P.

1926 The Melting Pot Mistake. Boston: Little Brown & Co.

Fletcher, Ronald

1991 Science, Ideology & the Media: the Cyril Burt affair. New Brunswick, NJ: Transaction Publishers

Freeman, Derek

1983 Margaret Mead and Samoa: the making and unmaking of an anthropological myth. Cambridge: Harvard University Press

Fustel de Coulanges, Denis

1864 The Ancient City. Translated from the 12th edition (1888). Paris: Hachette

Galton, Sir Francis

1869 Hereditary Genius: an enquiry into its laws and consequences. London: Macmillan

1889 Natural Inheritance. New York: Macmillan

1908 Memories of My Life. New York: Dutton

1909 Essays in Eugenics. London: Eugenics Education Society

Garrett, Henry E.

1960 "Klineberg's Chapter on Race and Psychology". *The Mankind Quarterly* 1: 1

Goldsmith, Timothy H.

1991 The Biological Roots of Human Nature: forging links between evolution and behavior. New York: Oxford University Press

Gould, Stephen J.

1981 The Mismeasure of Man. New York: W. W. Norton.

Graham, Robert Klark

1970 The Future of Man. With introduction by Sir Cyril Burt. North Quincy, MA: The Christopher Publishing House

Gramsci, Antonio

1977 Selections from Political Writings, 1910–1920. New York: International Publishing Co.

Grant, Madison

1904 Trail and Camp Fire. New York: Boone and Crockett Club

1918 The Passing of the Great Race: or the racial basis of European history. New York: Charles Scribner's Sons

1921 The Conquest of a Continent: or the expansion of races in America. New York: Charles Scribner's Sons.

Haeckel, Ernst

1905 The Evolution of Man. Trans. by Joseph McCabe of *Anthropogenie*, New York: D. Appleton

Haldane, J. B. S.

1924 Daedelus, or The Science of the Future. London: Kegan Paul

Hall, Ruth

1977 Passionate Crusader: the life of Marie Stopes. New York: Harcourt Brace and Jovanovich

Hardin, Garrett

1977 The Limits of Altruism: an ccologist's view of survival. Bloomington: The Indiana University Press.

1993 Living Within Limits: ecology, economics, and population taboos. New York: Oxford University Press

Hankins, Frank H.

1926 The Racial Basis of Civilization.

Herrnstein, Richard J.

1971 "I.Q". *Atlantic Monthly* (September): 43-46

1973 IQ in the Meritocracy. Boston: Atlantic-Little Brown

Herrnstein, Richard J. and Charles Murray

1994 The Bell Curve: intelligence and class structure in American life. New York: The Free Press

Herskovits, Melville
1953 Franz Boas. New York: Charles Scribner's Sons
Heuse, Georges A.
1955 "Race, Racismes, Antiracismes". *Revue de Psychologies des Peuples* 10: 3
Holden, C.
1994 "'Times' Corrects Scientist's Orbit". *Science* 263, 922
Hooton, Earnest Albert
1931 Up from the Ape. New York: Macmillan
1939 Apes, Men and Morons. New York: G.P. Putnam's Sons
1941 Why Men Behave Like Apes and Vice Versa. Princeton: Princeton University Press
Horgan, John
1993 "Eugenics Revisited". *Scientific American*, June 1993, pp 122-131.
Hrdlička, Aleš
1925 Old Americans. Baltimore: The Williams & Wilkins Company
1928 "Race Deterioration and Destruction with Special Reference to the American People". in The Proceedings of the Third Race Betterment Conference. Battle Creek, MI: Race Betterment Foundation
Hughes, Austin L.
1988 Evolution and Human Kinship. New York: Oxford University Press
Huntington, Ellsworth
1935 Tomorrow's Children. New York: John Wiley
Huxley, Sir Julian
1937 Foreword to *Eugenics and Politics* by Herbert Brewer. London: Eugenics Society
1941 Eugenics and Society, in *Man and the Modern World*. Reprinted 1944, London: Scientific Book Club
1947 Evolution and Ethics. London: Pilot
1951 Introduction to *Human Fertility* by Robert C. Cook, New York: Sloan
1962 "Eugenics in Evolutionary Perspective". *The Eugenics Review*, October. p. 133
Jaeger, Werner
1945 Paidea: The Ideals of Greek Culture. Trans. G. Highet. New York: Oxford University Press
Jamieson, J. W.
1982 "Conway Zirkle and the Persistence of 'Marxian biology' in the Western Social Sciences". *The Mankind Quarterly* 22: 3
1993 "Concerning Scholarly Creativity: Hermann J. Muller and

Germinal Repositories". *The Mankind Quarterly* 33: 4

Jennings, H. S
1925 Prometheus or Biology and the Advancement of Man. New York: E. P. Dutton and Co.
1930 The Biological Basis of Human Behavior. New York: W.W. Norton

Jensen, A. R.
1969 "How Much Can we Boost IQ and Scholarly Achievement?" *Harvard Education Review*. 39 (1)
1973 Educability and Group Differences. London: Methuen
1980 Bias in Mental Testing. New York: Free Press

Jones, Steven
1994 The Language of Genes: Solving the Mysteries of our Genetic Past, Present and Future. New York: Anchor Books

Jordan, David Starr
1915 War and the Breed. Repr. 1984. Washington D.C.: Cliveden Press

Josey, Charles C.
1922 The Social Philosophy of Instinct. New York: Charles Scribner's Sons
1923 Race and National Solidarity. Reprinted 1995 under the title of The Philosophy of Nationalism by Scott-Townsend Publishers, Washington D.C.

Joynson, Robert B.
1989 The Burt Affair. London: Routledge and Kegal Paul

Jürgens, Hans
1981 The Face of Africa. Washington D.C.: The Cliveden Press

Kamin, L. J.
1974 The Science and Politics of IQ. Potomac, MD: Erlbaum

Keith, Sir Arthur
1928 Concerning Man's Origin. London: Putnam's Sons
1940 A New Theory of Human Evolution, by Arthur Keith. London: Watts and Co.
1947 Evolution and Ethics. New York: Putnam
1955 Darwin Revalued. London: Watts & Co.

Kevles, Daniel J.
1985 In the Name of Eugenics: genetics and the use of human heredity. New York: Harper and Row

Klehr, Harvey
1988 Far Left of Center: the American radical Left today. New Brunswick: Transaction Books

Lappé, Marc A.
1994 "Justice and the Limitations of Genetic Knowledge" in *Justice*

and the Human Genome Project, eds. T.F. Murphy and Marc A. Lappé. Berkeley: California University Press

Lenski, Gerhard E.
1966 Power and Privilege: a theory of social stratification. New York: McGraw.

Levin, Michael
1994 "Comment on the Minnesota Transracial Adoption Study". *Intelligence*. 19: 13-20

Lewontin, R. C., S. Rose, and L. J. Kamin
1984 Not in our Genes; biological ideology and human nature. Pantheon, New York

Lynn, Richard
1991 "The Evolution of Racial Differences in Intelligence". *The Mankind Quarterly* 32:1-2

MacIver, R. M.
1921 The Elements of Social Science. New York: Dutton

Mannheim, Karl
1936 Ideology and Utopia: an introduction to the sociology of knowledge. New York: Harcourt Brace Jovanovich.

Mayr, Ernst
1963 Animal Species and Evolution. Cambridge: Harvard University Press.

McDougall, William
1908 An Introduction to Social Psychology. London: Methuen
1921 Is America Safe for Democracy? New York: Charles Scribner's Sons
1939 Group Mind. London: Cambridge University Press

Mead, Margaret
1928 Coming of Age in Samoa. New York: Morrow

Mead, Margaret, and Gene Weltfish
1940 The Races of Mankind. Pamphlet no. 85. New York: Public Affairs Committee, Inc.

Montagu, M. F. Ashley
1942 Man's Most Dangerous Myth: the fallacy of race. 2d ed. New York: Columbia University Press

Mourant, A.E.
1954 The Distribution of Human Blood Groups. Springfield, IL: Charles C. Thomas

Mourant, A.E., , A. C. Kopec and K. Domaniewska-Sobczak
1978 The Genetics of the Jews. Oxford: Clarendon Press

Muller, Hermann J.
1935 Out of the Night. New York: Vanguard.
1950 "Our Load of Mutations" *American Journal of Human Genetics,*

June Vol. 2 No. 2
1963 Human Genetic Betterment. New York: American Humanist Association.

Murphy, J.
1982 Evolution, Morality, and the Meaning of Life. Totowa NJ: Rowman and Littlefield

Murphy, Timothy F., and Marc A. Lappé (Eds.)
1994 Justice and the Human Genome Project. Berkeley: University of California Press

Osborn, Frederick
1968 The Future of Human Heredity: an introduction to eugenics in modern society. New York: Weybright and Talley

Palmer, L. R.
1955 Achaeans and Indo-Europeans. New York: Oxford University Press

Pauling, Linus
1959 "Molecular Disease". *Amer. Journal of Orthopsychiatry* 29: 4

Pearson, Karl
1892 The Grammar of Science. London: Scott
1901 National Life from the Standpoint of Science. London: Black
1903 "Huxley Memorial Lecture". *British Medical Journal*, Oct. 10
1909 The Scope and Importance to the State of the Science of National Eugenics. London: The Galton Eugenics Laboratory Lectures

Pearson, Roger
1974 Introduction to Anthropology. New York: Holt, Rinehart and Winston
1985 Anthropological Glossary. Malabar, FL: Robert E. Krieger Publishing Company
1991 Race, Intelligence and Bias in Academe. With an introduction by Hans J. Eysenck. Washington, D.C.: Scott-Townsend Publishers
1992 Shockley on Eugenics and Race. With an introduction by Arthur R. Jensen. Washington, D.C.: Scott-Townsend Publishers

Phillpotts, Bertha S.
1913 Kindred and Clan in the Middle Ages and After. Cambridge: Cambridge University Press

Pressley, Sue Anne
1995 "After Years of Birth Defects, Pain and Mystery Linger in Brownsville". *Washington Post*, September 17

Price, Joyce
1995 "Crime-and-Genes Conference Resurfaces with New Focus".

Washington Post. September 22

Ross, Edward A.
1923 The Outlines of Sociology. New York: The Century Co.
1927 Standing Room Only! New York: The Century Co.

Rowe, J. Stan
1991 Summing it Up. in *Planet Under Stress: The Challenge of Global Change.* Constance Mungall and Digby J. McLaren, Editors. New York: Oxford University Press

Rushton, J. Philippe
1989 "Genetic Similarity, Human Altruism, and Group Selection". *Behavioral and Brain Sciences* 12: 503-559
1992 "Cranial Capacity Related to Sex, Rank and Race in a Stratified Random Sample of 6,325 U.S. Military Personnel", *Intelligence* 16: 401-413
1995 Race, Evolution and Behavior: a life history perspective. New Brunswick: Transaction Publishers

Safri, Ahmad
1992 Max Weber's Sociology of Intellectuals. London: Oxford University Press

Saitou, N
1995 "A Genetic Affinity Analysis of Human Populations". *Human Evolution* 10:1, 17-33

Schiff, M. and R. Lewontin
1986 Education and Class: the irrelevance of IQ genetic studies. Oxford: Clarendon Press

Segal, Nancy
1984 "The Nature vs. Nurture Laboratory". *Research* July/August

Seligman, Daniel
1992 A Question of Intelligence; the IQ debate in America. Secaucus NJ: Birch Lane Press

Shaw, George Bernard
1965 Preface to Man and Superman, in *The Complete Works of Bernard Shaw.* London: Paul Hamlyn

Shipman, Pat
1994 The Evolution of Racism. New York: Simon and Shuster

Simpson, G. G.
1964 This View of Life: the world of an evolutionist. New York: Harcourt, Brace and World
1971 The Meaning of Evolution. New York: Bantam BooksStevens,
1980 Splendid Isolation. New Haven: Yale University Press

Snyderman, Mark, and Stanley Rothman
1988 The IQ Controversy: the media and public policy. New

Brunswick, NJ: Transaction Books.

Stevens, William K.

1994 "Evolution of Humans May at Last Be Faltering". *New York Times*, March 14, 1994.

Taylor, Jared

1992 Paved with Good Intentions. New York: Carroll and Graf Publishers

Tellegen, Auke, David T. Lykken, Thomas J. Bouchard, Jr., Kimerly J. Wilcox, Nancy L. Segal, Stephen Rich

1988 "Personality Similarity in Twins Reared Apart and Together." *Journal of Personality and Social Psychology*. 54: 6

Terman, Lewis M.

1926-30 Genetic Studies of Genius. 4 Vols. Stanford: Stanford University Press

Thomay, Laszlo F.

1993 The Natural Law of Race Relations. Washington DC: Scott-Townsend Publishers

Vallois, Henri

1951 "UNESCO on Race". *Man*, no. 28

van den Haag, Ernest

1963 "Social Science Testimony in the Desegregation Cases: A reply to Professor Kenneth Clark." *Villanova Law Review* 6: 1

Veblen, Thorstein

1914 The Instinct of Workmanship, New York: Macmillan

Waddington, C. H.

1975 The Evolution of an Evolutionist. Ithaca: Cornell University Press.

1978 The Man-Made Future. London: Croom Helm

Ward, Lester

[1883] Dynamic Sociology. Reprint, 1911. New York: Appleton & Co.

1899 Outlines of Sociology. New York: Macmillan

Weber, Max

1978 Economy and Society Vol. I, Translation edited by G. Roth and C. Wittich. Berkeley: University of California Press

1958 The Religion of India. (Ed. and translated by Hans M. Gert and Don Martindale). Glencoe, IL: Free Press

Weyl, Nathaniel

1989 The Geography of American Achievement. Washington: Scott-Townsend Publishers

Whetham, W.C.D. & C.D. Whetham

1909 The Family and the Nation: a study in natural inheritance and family responsibility. London: Longmans Green

Whitney, Glayde
1995 "Ideology and Censorship in Behavior Genetics". *The Mankind Quarterly* 35: 4

Wilkie, Tom
1993 Perilous Knowledge: the human genome project and its implications. Berkeley: University of California Press

Wilson, E. O.
1978 Sociobiology: the new synthesis. Cambridge: Harvard University Press

Wilson, J. Q. and R. J. Herrnstein
1985 Crime and Human Nature. New York: Simon and Schuster

Wolpoff, Milford
1989 "Multiregional Evolution: the fossil alternative to Eden" in *Behavioral and Biological Perspectives on the Origins of Modern Humans*. Eds. F.H. Smith and F. Spencer. New York: Alan R. Liss

Yerkes, R. M.
1921 "Psychological Examining in the United States Army". *Memoirs of the National Academy of Science.* 15

Zirkle, Conway
1949 Death of a Science in Russia: the fate of genetics as described in Pravda and elsewhere. Philadelphia: University of Pennsylvania Press
1959 Evolution, Marxian Biology, and the Social Scene. Philadelphia: University of Pennsylvania Press

INDEX